EMBEDDED SYSTEMS – MODELING, TECHNOLOGY, AND APPLICATIONS

Embedded Systems – Modeling, Technology, and Applications

Proceedings of the 7th International Workshop held at Technische Universität Berlin, June 26/27, 2006

Edited by

GÜNTER HOMMEL
Technische Universität Berlin
Berlin, Germany

and

SHENG HUANYE
Shanghai Jiao Tong University
Shanghai, China

Springer

A C.I.P. Catalogue record for this book is available from the Library of Congress.

ISBN-13 978-90-481-7230-6
ISBN-10 1-4020-4933-1 (e-book)
ISBN-13 978-1-4020-4933-0 (e-book)

Published by Springer,
P.O. Box 17, 3300 AA Dordrecht, The Netherlands.

www.springer.com

Printed on acid-free paper

Contents

Committee

Workshop Co-Chairs

Günter Hommel
Sheng Huanye

Program Committee

Günter Hommel
Sheng Huanye
Zhang Liqing

Organizing Committee

Wolfgang Brandenburg
Michael Knoke

Editorial Office and Layout

Wolfgang Brandenburg

Workshop Secretary

Anja Bewersdorff
Gudrun Pourshirazi

TU Berlin
Institut für Technische Informatik
und Mikroelektronik
Sekr. EN 10
Einsteinufer 17
10587 Berlin
Germany

Preface

The International Workshop on "Embedded Systems - Modeling, Technology, and Applications" is the seventh in a successful series of workshops that were established by Shanghai Jiao Tong University and Technische Universität Berlin. The goal of those workshops is to bring together researchers from both universities in order to present research results to an international community.

The series of workshops started in 1990 with the International Workshop on Artificial Intelligence and was continued with the International Workshop on "Advanced Software Technology" in 1994. Both workshops have been hosted by Shanghai Jiao Tong University. In 1998 the third workshop took place in Berlin. This International Workshop on "Communication Based Systems" was essentially based on results from the Graduiertenkolleg on Communication Based Systems that was funded by the German Research Society (DFG) from 1991 to 2000. The fourth International Workshop on "Robotics and its Applications" was held in Shanghai in 2000. The fifth International Workshop on "The Internet Challenge: Technology and Applications" was hosted by TU Berlin in 2002. The sixth International Workshop on "Human Interaction with Machines" was hosted by Shanghai Jiao Tong University.

The subject of this year's workshop has been chosen because the field of Embedded Systems has not only gained major interest in the research community but has also significant economic impact in different application fields. Mechanic, hydraulic, and electronic control systems are being replaced by microcomputer based embedded systems. In automotive and aircraft industries this has successfully been illustrated developing e.g. drive-by-wire and fly-by wire systems. The number of embedded microcontrollers in an Airbus rose to more than 500 and even in high-classed cars one can find more than 100 microcomputers.

The application of information technology in many devices requires not only the development of appropriate effective and efficient hardware components but also of dedicated software responsible for system control that additionally has to conform to high safety requirements. As an example think of robots or medical devices that should be used as helpmates directly in contact with handicapped or elderly persons.

All aspects from specification to modeling and implementation of hardware and software components for embedded systems are addressed in this workshop. Using those methods and tools successful realizations of embedded systems are described.

The continuous support of both universities is gratefully recognized that enabled the permanent exchange of ideas between researchers of our two universities.

Berlin, June 2006

Günter Hommel and Sheng Huanye

A Conceptual Model for Conformance, Compliance and Consistency

Sebastian Bab and Bernd Mahr

Formal Models, Logic and Programming
Technische Universität Berlin, Germany
{bab,mahr}@cs.tu-berlin.de

Summary. Theories of specification are built on assumptions on the processes of development. Theories which allow for specifications from different viewpoints raise the problem of nontrivial concepts of conformance, compliance and consistency. We propose a general conceptual model for these concepts, which applies to situations with multiple specifications prescribing requirements of different nature.

1 Concepts of Development in Theories of Specification

Development of distributed and embedded systems implies that one has to deal with multiple viewpoints, different types of requirements and various forms of description. As a consequence, at the level of system specification, there occur situations in which declarations concern different ontologies and requirements have their interpretation in different logics. This linguistic and at the same time conceptual heterogeneity can not remain completely unresolved. Since, by the nature of a development task, the targeted system is being developed as a functional whole, there is the need for unifying concepts of conformance, compliance and consistency. Not only is any judgement on the systems quality depending on such concepts, also from a pragmatic point of view these concepts are indispensable, since diversity of specifications is a major source of inefficiency and failure. It is therefore only natural to ask for specification theories which deal with conformance, compliance and consistency. Amazingly there is little systematic study of these concepts, which is independent from the languages and formalisms applied.

With the discipline of software engineering different theories of specification have emerged which address the general relationships between *requirements*, *specifications* and *realizations*. The first such theory views *realizations* as programs and *requirements* as problems to be algorithmically solved or as functions to be efficiently implemented. In this theory an answer to the question of conformance is given by proofs, which prove the

Günter Hommel and Sheng Huanye (eds.), Embedded Systems Modeling – Technology, and Applications, 1–10.

algorithm and the program which it implements correct. Compliance and consistency do not have a general meaning in this setting.

Since the mid-seventies the mathematics of abstract data types evolved into formal theories of algebraic specification, which view *realizations* as programs and *requirements* as properties of functionality. They assume that a realization is created in a process of stepwise refinement of structured *specifications* which all meet the initially stated requirements (see for example [2, 3 and 10]). In these theories conformance, and accordingly compliance, is achieved by properly executing steps of refinement. Consistency is on the one hand guaranteed by the restricted form of specifications, like, for example, equational specifications which are free from any inconsistencies, and is, on the other hand, proven to be implied if specifications are being derived by correctly applied constructors, like parameter passing and combination.

On the basis of foundational mathematical theories, like set theory and category theory, various specification languages and frameworks have been developed. They usually provide means for horizontal and vertical structuring in modular designs (see [4]). Due to their abstract concepts of semantics, such as institutions and general logics, these languages and frameworks have laid the ground for the integration of formalisms as a means to deal with different aspects of specification in a consistent way (see [7]). The first such approach studied the integration of data type specifications and process descriptions (see [5]). Later approaches treated integration in more general terms (see [9]). In all these conceptualizations conformance, compliance and consistency are being treated in an axiomatic manner and in a way similar to that of abstract data type oriented specification languages.

Since the mid-eighties object orientation in systems analysis, design, programming and architecture allowed to effectively dealing with distribution and heterogeneity by means of encapsulation, interfacing and communication. Object oriented techniques brought with it a change in the general perception. Objects were seen as models rather than specifications. As a consequence the concepts of requirements, specifications and realizations as well the questions of conformance, compliance and consistency lost the theoretical and mathematical basis of their meaning. If not abandoned, they were treated by means similar to those of program correctness, like for example in Eiffel, or were just delegated to reuse and standardization.

And indeed, international standardization has linked the concepts of the mathematically founded specification frameworks with the concepts of object orientation and with best practice experience in architecting. An important outcome of these activities is the reference model for open distributed processing (RM-ODP), issued by the ISO in the mid-nineties.

Together with its associated standards the RM-ODP is best understood as a general specification theory for open distributed systems and applications (see [6]).

Implicit in the conceptualizations of the RM-ODP is the following theory of development: *requirements* concern the use and operation of the targeted system in its environment. The targeted system itself is considered to be a *realization* of interrelated *specifications*. Specifications are classified by viewpoints and are being developed in appropriate steps of refinement and translation. At the same time the process of specification development is being supported by the use of tools and by the use of standards constraining specifications.

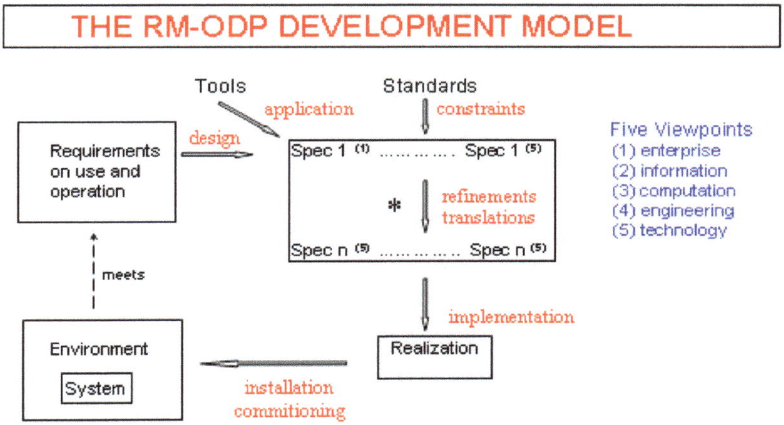

The RM-ODP explicitly defines the concepts of conformance, compliance and consistency: a realization R is said to be *conformant* with respect to a specification S, if R meets the requirements prescribed by S; a specification S_2 is said to be *compliant* with a specification S_1 if every realization which is conformant with S_2 is also conformant with S_1; and two specifications are said to be *consistent*, if there is a realization which is conformant to both. While compliance and consistency of specifications under one and the same viewpoint can be thought of being enforced by proper steps of refinement and translation, the idea of compliance and consistency of specifications under different viewpoints creates a conceptual problem not properly covered by the reference model.

Being derived from different techniques for analysis and design UML has become an internationally standardized language and tool for description (see [8]). It provides several diagrammatic techniques for the modelling of functionality, architecture, interrelation and behaviour of systems and their components. Questions of conformance are outside the scope of UML. They are being addressed in theoretical studies on UML semantics and in tools automatically deriving programs from UML models. The conceptualization of consistency of specifications using different diagrammatic techniques has become a matter of research in recent years and remains widely unsolved (see [12, 13]).

In the following sections a general model is defined which in terms of abstract logic conceptualizes specifications whose prescriptions may concern different ontologies and different kinds of constraints. Based on this unifying conceptualization of specifications the concepts of conformance, compliance and consistency are then defined in their classical form. A source for this conceptualization is also [11] where the concept of reference points for the observation of systems behaviour has been investigated.

2 The Logical Basis of Specifications

Specifications prescribe requirements. Requirements are expressed and interpreted on a given logical basis. A logical basis consists of three given parts: a *domain of discourse*, a *domain of descriptions*, and *interpretational relationships* relating the two domains. More specifically, a logical basis is given as a triple = (***dis***, ***des***, ***int***), where:

- ***dis*** is called the *domain of discourse* and consists of:

  a set of *entities*, denoted by

  entity

  and a binary *membership* relation

  is_element_of: <entity,entity>

  together with the derived *ownership* relation

  belongs_to: <entity,entity>

  subject to the condition

  e_1 belongs_to e_2 iff e_1 is_element_of e_2 or there exists an entity x such that e_1 belongs_to x and x is_element_of e_2.

  and the derived *part of* relation

  is_part_of: <entity,entitiy>

  subject to the condition

  e_1 is_part_of e_2 iff x belongs_to e_1 implies x belongs_to e_2 for all entities x.

The set **entity** in the domain of discourse contains the things to which specifications refer by their declarations and propositions. The membership relation **is_element_of** associates with each entity e those entities, of which e is directly composed. The ownership relation **belongs_to** describes iterated ownership while the relation **is_part_of** describes iterated subsets. Note that the membership relation **is_element_of** may contain reflexive pairs while **belongs_to** and **is_part_of** are irreflexive.

- ***des*** is called the *domain of descriptions* and consists of:

a set of (well-formed) *expressions*
expression
a set of (well-formed) *types*
type
a set of (well-formed) *sentences*
sentence
a set of (well-formed) *propositions*
proposition
where a proposition p has either of the forms
φ:true
φ:false
with **φ** being a sentence
and a set of (well-formed) *declarations*
declaration
where a declaration d has the form
e:t
with **e** being an expression and **t** being a type.
For reasons of abbreviation we define the set of (well-formed) statements
statement
by **statement** := **declaration** ∪ **proposition**.

- ***int*** is called *interpretational relationships* and consists of:

a pair of *meaning* functions
[[ ]]: expression → entitiy
[[ ]]: type → entity
and a *satisfaction* relation
⊨ : entity → sets_of(statement)

subject to the conditions

A ⊨ e:t	**iff**	**[[t]] is_part_of A and [[e]] is_element_of [[t]]**
A ⊨ φ:true	**iff**	**not A ⊨ φ:false**
A ⊨ Σ	**iff**	**A ⊨ σ for all statements σ in Σ.**

While declarations require entities of a certain type to exist, propositions require entities to have certain properties. This concept of logical basis combines first order interpretation of ∈-logic, disciplines of declaration and abstract logics (see [1]).

3 Specifications

Specifications prescribe requirements on a given logical basis. Requirements may be obligations, prohibitions or permissions, which are expressed as certain modalities of assertions.

Specifically, given a logical basis **B** = (***dis***, ***des***, ***int***),

- the set of (well-formed) *assertions* is denoted by

  <u>assertion</u>

  and contains all pairs of the form

  (e,Σ)

  with **e** being an expression and

  Σ being a set of statements.

  We say that an assertion **(e,Σ)** is *valid* and write

  (e,Σ):valid

  if and only if

  [[e]] ⊨ σ for all σ in Σ

  and say that an assertion **(e,Σ)** is *invalid* and write

  (e,Σ):invalid

  if and only if

  [[e]] ⊨ σ does not hold for some σ in Σ.

- the set of (well-formed) *modalities* is denoted by

  modality

  and is defined by

  modality := **fact** ∪ **knowledge** ∪ **observability**

  where

  fact

  denotes the set of *factual assertions* having the form **α:fact**, with **α** being an assertion,

  knowledge

  denotes the set of *known assertions* having the form **α:known**, with **α** being an assertion, and

  observability

  denotes the set of *observable assertions* having the form **α:observable**, with **α** being an assertion.

- the set of (well-formed) *requirements* is denoted by

  requirement

  and is defined by

  requirement:=**obligation** ∪ **prohibition** ∪ **permission**

  where

  obligation

  is the set of *modalities being obliged* having the form **μ:obliged**, with **μ** being a modality,

  prohibition

  is the set of *modalities being prohibited* having the form **μ:prohibited**, with **μ** being a modality,

  permission

  is the set of *modalities being permitted* having the form **μ:permitted**, with **μ** being a modality.

- the set of *specifications* is denoted by

  specification

  and associated with a function

  prescription: specification → sets_of(requirement).

Note that there are no constraints on the choice of requirements. Inconsistencies in prescriptions are being detected by the judgement of conformance.

4 Conformance, Compliance and Consistency

The concept of conformance expresses the fact that an object is conformant to a specification. This is to say that an object under consideration fulfils the requirements prescribed by its specification. Objects are models representing real world situations in which certain assertions are a fact, certain assertions are known and certain assertions are observable. This view on objects and on their specification motivates the conceptualization of conformance.

Specifically, given a logical basis = (***dis***, ***des***, ***int***),

- the set of *objects* is denoted by

  <u>object</u>

  where an object is a combination

  obj = (F,K,O)

  consisting of sets **F**,**K** and **O** of assertions such that

  $\mathbf{K} \subseteq \mathbf{F}$
  $\mathbf{K} \subseteq \mathbf{O}$
  (e, Σ) in F implies (e, Σ):valid.

- Given a specification **S** and an object **obj**, we say that **obj** *is conformant with* **S**, written

  obj conforms_to S

  if and only if **obj** fulfils all requirements prescribed by **S**.

- We say that an object **obj** *fulfils a requirement* **r**, written

  obj $\models$ **r**

  if and only if

  μ is the case for **obj**, if **r** equals **μ:obliged**
  μ is not the case for **obj**, if **r** equals **μ:prohibited**
  μ is the case for **obj** or **μ** is not the case for **obj,**
  if **r** equals **μ:permitted**,

  where a modality **μ** is said to *be the case for* an object **obj, written**

  obj $\models$ **μ**

  if and only if

  α is in **F,** if **μ** is of the form **α:fact**
  α is in **K,** if **μ** is of the form **α:known**
  α is in **O,** if **μ** is of the form **α:observable**.

- A specification $\mathbf{S_2}$ *is compliant with* a specification $\mathbf{S_1}$, written
  $\mathbf{S_2}$ complies_to $\mathbf{S_1}$
  if and only if every object which is conformant with $\mathbf{S_1}$ is also conformant with $\mathbf{S_2}$.

- A specification $\mathbf{S_1}$ *is consistent with* a specification $\mathbf{S_2}$, written
  $\mathbf{S_1}$ consistent_with $\mathbf{S_2}$
  if and only if there is an object **obj** which is conformant with both, $\mathbf{S_1}$ and $\mathbf{S_2}$. Note, that this definition allows $\mathbf{S_1}$ and $\mathbf{S_2}$ to be equal.

5 Conclusions

In this conceptual model for conformance, compliance and consistency we have abstracted from the form of specifications and from the techniques of their interrelation. Instead, the model focuses on the ways to deal with the diversity of requirements. It is only a first proposal. As such it remains to be studied further and to be applied to real situations of development.

References

1. J.P. Cleave. *A Study of Logics*. Clarendon Press, 1991.
2. Hartmut Ehrig, Hans-Jörg Kreowski, Bernd Mahr, and Peter Padawitz. Algebraic implementation of abstract data types. *Theoretical Computer Science*, 20:209-263, 1982.
3. Hartmut Ehrig and Bernd Mahr. *Fundamentals of Algebraic Specification 1*. Springer Verlag Berlin, 1985.
4. Hartmut Ehrig and Bernd Mahr. *Fundamentals of Algebraic Specification 2*. Springer Verlag Berlin, 1990.
5. Hartmut Ehrig and Fernando Orejas. A generic component concept for integrated data type and process modeling techniques. Technical Report 2001/12, Technische Universität Berlin , 2001.
6. International Organization for Standardization. *Basic Reference Model of Open Distributed Processing*. ITU-T X.900 series and ISO/IEC 10746 series, 1995.
7. Joseph A. Goguen and Rod M. Burstall. Institutions: abstract model theory for specification and programming. *Journal of the ACM*, 39(1): 95-146, 1992.
8. OMG.ORG: Object Management Group. Download of the UML Specification. 2005.
9. T. Mossakowski. Foundations of heterogeneous specification. In M. Wirsing, D. Pattinson, and R. Hennicker, editors, *Recent Threads in*

Algebraic Development Techniques, 16th International Workshop, WADT 2002, pages 359-375. Springer London, 2003.

10. Martin Wirsing. Algebraic Specification. In *Handbook of Theoretical Computer Science, Volume B: Formal Models and Semantics (B)*, pages 675-788. 1990.
11. Sebastian Bab and Bernd Mahr. Reference Points for the Observation of Systems Behaviour. In Hommel, Huanye, editors, *Human Interaction with Machines*, Springer, 2006.
12. Workshop on "Consistency Problems in UML-based Software Development". In *UML 2002: Model Engineering, Concepts and Tools*. Blekinge Institute of Technology, Research Report 2002:06.
13. Workshop on "Consistency Problems in UML-based Software Development II". In *UML 2003: Modeling Languages and Applications*, October 20, 2003, San Francisco, USA. Blekinge Institute of Technology, Research Report 2003:06.

Simulation and Animation of Visual Models of Embedded Systems

A Graph-Transformation-Based Approach Applied to Petri Nets

Hartmut Ehrig, Claudia Ermel, and Gabriele Taentzer

Theoretical Computer Science – Formal Specification Techniques
Technische Universität Berlin, Germany
{ehrig,lieske,gabi}@cs.tu-berlin.de

Summary. Behavior specification techniques like Petri nets provide a visual description of software and embedded systems as basis for behavior validation by simulation. Graph transformation systems can be used as a unifying formal approach to define various visual behavior modeling languages including different kinds of Petri nets, activity diagrams, Statecharts etc., and to provide models with an operational semantics defining simulations of visual models based on graph transformation rules. Moreover, simulation of visual models can be extended by animation which allows to visualize the states of a model simulation run in a domain-specific layout which is closer to the problem domain than the layout of the abstract diagrammatic notation of the specification technique. This kind of model transformation is defined also within the framework of graph transformation, which allows to show interesting properties like semantical correctness of the animation with respect to simulation. In this paper we give an overview of simulation and animation of visual models based on graph transformation and discuss corresponding correctness issues.

As running example we use a high-level Petri net modeling the basic behavior of an elevator. We show how Petri nets are mapped to graph transformation systems, and how the elevator system is extended using an animation view which shows the movements of an elevator cabin between different floors.

1 Introduction

Visual modeling techniques provide an intuitive, yet precise way in order to express and reason about concepts at their natural level of abstraction. The success of visual techniques in computer science and engineering resulted in a variety of methods and notations addressing different application domains and different phases of the development process.

Nowadays two main approaches to visual language definition can be distinguished: grammar-based approaches or meta-modeling. Using graph grammars, visual sentences are described by graphs and visual languages by graph

Günter Hommel and Sheng Huanye (eds.), Embedded Systems Modeling – Technology, and Applications, 11–20.

grammars. Meta-modeling is also graph-based, but uses constraints instead of a grammar to define the visual language. While visual language definition by graph grammars can borrow a number of concepts from classical textual language definition, this is not true for meta-modeling.

In this paper, we apply the graph grammar-approach to visual language definition [1]. The concrete as well as the abstract syntax of visual notations are described by typed attributed graphs. The type information given in the type graph, captures the definition of the underlying visual alphabet, i.e. the symbols and relations which are available. A visual language is defined by an alphabet and a typed attributed graph grammar, the *VL syntax grammar.*

The behavior of visual models (e.g. Petri nets or Statecharts) can be described by graph transformation as well. Graph rules called *simulation rules* are used to define the operational semantics of the model. Simulation means to show the before- and after-states of an action as diagrams. In Petri nets, for example, a simulation is performed by playing the token game. Different system states are modeled by different markings, and a scenario is determined by a firing sequence of transitions resulting in a sequence of markings. In the formal framework of graph transformation, a scenario corresponds to a transformation sequence where the simulation rules are applied to the graphs representing system states.

However, for validation purposes simulation is not always adequate, since system states are visualized in simulation runs as diagrams. On the one hand, these state graphs abstract from the underlying system domain (as modeling VLs serve many purposes), on the other hand they may become rather complex as complete system states involve auxiliary data and constructs which are necessary to control the behavior but which do not provide insight in the functional behavior of the specific model.

In this paper we extend the simulation for a visual model by *animation views* which allow to define domain-specific scenario animations in the layout of the application domain (cf. [7]). We call the simulation steps of a behavioral model *animation steps* when the states before and after a step are shown in the animation view.

To define the animation view, the VL alphabet first is extended to an animation alphabet by adding symbols representing the problem domain. Second, the set of simulation rules for a specific visual model is extended, resulting in a new set of graph transformation rules, called *animation rules.*

The paper is organized as follows: In Section 2 we introduce our running example, namely the basic behavior of an elevator modeled as high-level Petri net, and present a domain specific animation view of the system as motivation for the further sections. In Section 3 we define the syntax of visual languages and simulation of visual models, and discuss the correctness of simulation. The extension from simulation to animation is presented in Section 4 together with a discussion of the correctness of animation. In the conclusion in Section 5 we discuss related work and implementation issues.

2 Example: An Elevator Modeled as Petri Net

Petri nets are a standard modeling technique for reactive and concurrent systems. In our running example, we use high-level Petri nets to model the basic behavior of an elevator which could be easily extended to a system of concurrent elevators.

In high-level Petri nets, structured tokens are used to represent the data of a system, whereas the Petri net captures the modification of the data. An *Algebraic High-Level net* (AHL net for short) [8] consists of a Place/Transition net for the process description and an algebraic specification $SPEC$ for the data type description describing operations used as arc inscriptions. Tokens are elements of a corresponding $SPEC$-algebra. In order to keep the example simple, the AHL net does not model the control mechanism to call the elevator but only the movements up and down (see Fig. 1 (a)). In general, the system states show in which floor the elevator cabin is, and whether the elevator is moving or not. The data type specification $SPEC$ consists of the specification Nat of natural numbers and the two constants $MaxFloor = 4$ and $MinFloor = 0$. Thus, in Fig. 1 (a), a house with five floors is modeled. The variable f used as arc inscription is holding the number of the current floor of the elevator cabin. The initial marking specifies that in the beginning the elevator is in the ground floor, in the state not moving (token 0 on place not moving).

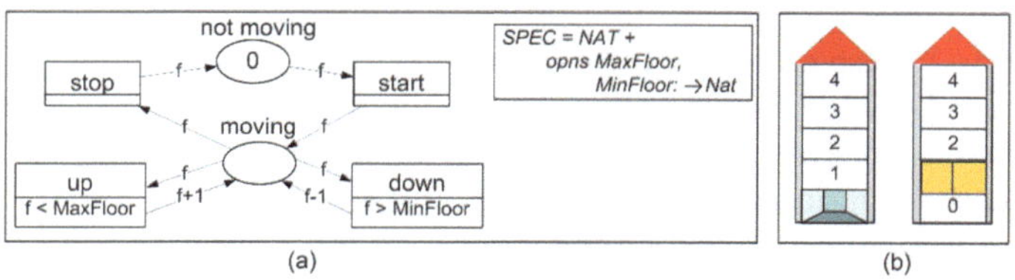

Fig. 1. A Basic Elevator as AHL Net (a) and Snapshots of its Animation View (b)

A domain-specific animation view of the AHL net is illustrated in Fig. 1 (b). We show two snapshots of the animation view according to two possible markings of the net in Fig. 1. The elevator is illustrated as part of a building showing the actual number of floors. The elevator cabin is visualized as box with doors when moving, and as an open box when the elevator is standing, The state not moving is visualized in the left snapshot of Fig. 1 (b), where the cabin is positioned in the respective floor and the doors are open. This snapshot corresponds to the initial state of the AHL net shown in Fig. 1 (a). When the elevator is moving between floors (state moving), the cabin doors are closed. This is shown in the right snapshot of Fig. 1 (b), and corresponds to a token "1" on place moving. We will ensure that the actions that can be performed in the animation view correspond to the transitions in the AHL net.

3 Visual Languages and Simulation of Visual Models

We start with a short introduction to graph transformation which is our basis for modeling visual languages.

3.1 Graph Transformation

The main idea of graph grammars and graph transformation is the rule-based modification of graphs where each application of a graph transformation rule leads to a graph transformation step. Graph grammars can be used on the one hand to generate graph languages similar to Chomsky grammars in formal language theory. On the other hand, graphs can be used to model the states of all kinds of systems and graph transformation to model state changes.

Especially for the application of graph transformation techniques to visual language (VL) modeling, *typed attributed graph transformation systems* have proven to be an adequate formalism [3, 4].

The core of a graph transformation rule $p = (L, R)$ is a pair of graphs called left-hand side L and right-hand side R, and an injective (partial) graph morphism $r : L \rightarrow R$. Intuitively, the application of rule p to graph G via a match m from L to G deletes the image $m(L)$ from G and replaces it by a copy of the right-hand side R. A typed graph grammar $GG = (TG, P, S)$ consists of a type graph TG, a set of rules P, and a start graph S. The language of a graph grammar consists of the graphs that can be derived from the start graph by applying the transformation rules arbitrarily often.

Moreover, we use node attributes in our examples, e.g. text for the names of nodes, or numbers for their positions. This allows us to perform computations on attributes in our rules and offers a powerful modeling approach. For flexible rule application, variables for attributes can be used, which are instantiated by concrete values in the rule match.

An example for a graph grammar with node attributes is the visual syntax grammar for AHL nets which is explained in Section 3.2. For a formal theory of typed attributed graph transformation systems we refer to [3].

3.2 Syntax of Visual Languages

A visual language is specified by an alphabet for symbol and link types (the type graph), and a grammar consisting of a start sentence and a set of syntax rules. A visual language VL is then the set of all visual sentences that can be derived by applying syntax rules. The abstract syntax of the alphabet for the AHL net language is shown in Fig. 2 (a).

At the abstract syntax level, there are symbols as Place, Transition, etc. in Fig. 2 which are attributed by datatypes like PlName of type String or Value of type Nat. These abstract syntax items are linked by directed edges. We distinguish arcs running from places to transitions (type PreArc) and arcs from transitions to places (type PostArc). The concrete syntax level is shown

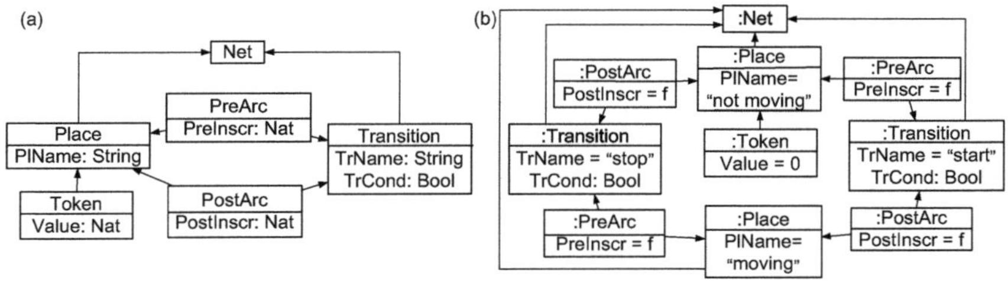

Fig. 2. Abstract Syntax of Visual AHL Net Alphabet (a) and Visual Sentence (b)

in Fig. 7. An example for a visual sentence in its abstract syntax is depicted in Fig. 2 (b). It shows a part of the abstract syntax of the elevator net in Fig. 1.

Syntax rules for AHL nets define the generation of places and transitions, and of arcs between them. A sample syntax rule insPreArc(i) the abstract syntax of which is depicted in Fig. 3, defines the operation for inserting an arc from a place to a transition. Note that graph objects which are preserved by the rule (the place and the transition), occur in both L and R (indicated by equal numbers for the same objects). The newly created arc is inserted together with the arc inscription defined by the rule parameter i (see [6] for more details.) A suitable rule application control has to ensure that rule insPreArc(i) is applied only once for a given place and transition.

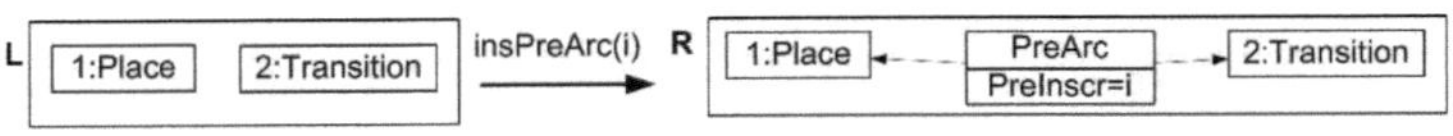

Fig. 3. Sample Abstract Syntax Rule from the AHL Net Syntax Grammar

3.3 Simulation of Visual Models

As discussed already in the introduction, we can use graph grammars not only to define the syntax of visual languages, but also for the simulation of visual models. A visual model is given by the subset of all visual sentences of a given visual language, which correspond to a specific model, like the AHL net model of our elevator in Fig. 1. The behavior of the AHL net in Fig. 1 is given by the well-known token game of AHL nets (see [8]).

In our framework the behavior of a visual model is defined by graph transformation rules based on the abstract syntax of the corresponding visual sentences. These rules are called *simulation rules* in contrast to the syntax rules of the visual language. In order to define the simulation rules we use a suitable encoding of Petri nets into graph grammars (see also [11]). A very natural encoding of nets into grammars regards a net as a graph grammar acting on graphs representing the markings of a net. A Petri net transition is represented as a graph transformation rule consuming the tokens in the pre domain and generating the tokens in the post domain of the transition. Places are represented as graph nodes which are preserved by the rule. Fig. 4 shows the schema for the translation of an arbitrary AHL net transition to a graph transformation rule $L \to R$, where L holds the pre domain tokens to be consumed and R contains the post domain tokens to be generated.

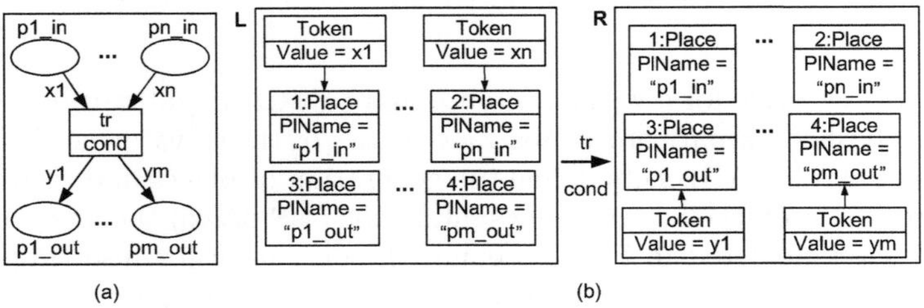

Fig. 4. (a) AHL Net Transition, (b) Corresponding Graph Transformation Rule

For our elevator net in Fig. 1, we have four simulation rules, one for each transition of the net. Three of them are shown in Fig. 5 (the rule stop equals the inverse rule of rule start).

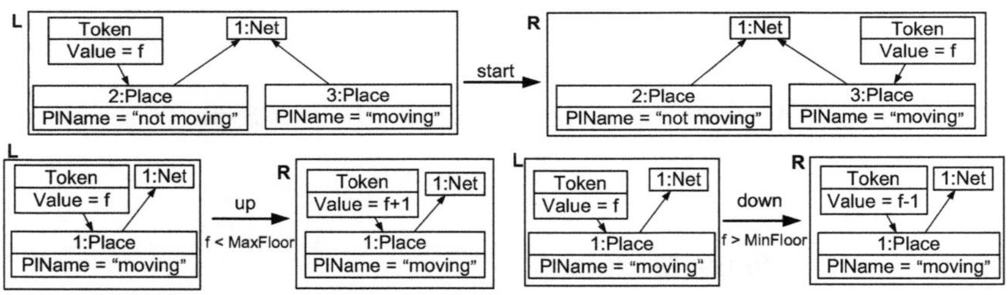

Fig. 5. Simulation Rules for the Elevator Model

If the visual model is given by a specification using a formal specification technique like Petri nets, which have already a well-defined operational semantics, it is important to show that the visual modeling and simulation using graph grammars is correct w.r.t. the given formal specification technique. Otherwise, it is problematic to apply formal analysis and verification techniques. In our case, the formal relationship between AHL nets and graph grammars has been analyzed in [8], where the semantical compatibility of AHL nets and the corresponding attributed graph grammar with simulation rules has been shown as one of the main results.

4 From Simulation to Animation of Visual Models

In order to bridge the gap between the underlying formal, descriptive specification as a Petri net and a domain-specific dynamic visual representation of processes being simulated, we suggest the definition of animation views for Petri nets. Of course, the behavior shown in the animation view has to correspond to the behavior defined in the original Petri net.

4.1 Animation of Visual Models

The elevator net as illustrated in Fig. 1 (a), is a visual sentence of our AHL net language. The animation view of this sentence has been already motivated by Fig. 1 (b), where an elevator cabin moving between the floors of a house is visualized. The concrete and abstract syntax of the alphabet of the animation view is depicted in Fig. 6. It contains a symbol type for the elevator shaft (House), types for the different states of the elevator cabin (Open Cabin and LeftDoor and RightDoor, which are positioned within the corresponding symbol of type Floor). The concrete syntax both for the AHL net elements and for the new elements of the animation view is described by further graphic nodes and by graphical constraints between them specifying the intended layout. We use dotted arrows at the concrete syntax level in order to indicate the graphical constraints, e.g. that a token value is always written *inside* the corresponding Place ellipse, or that the left elevator door is positioned in the left part of the floor rectangle.

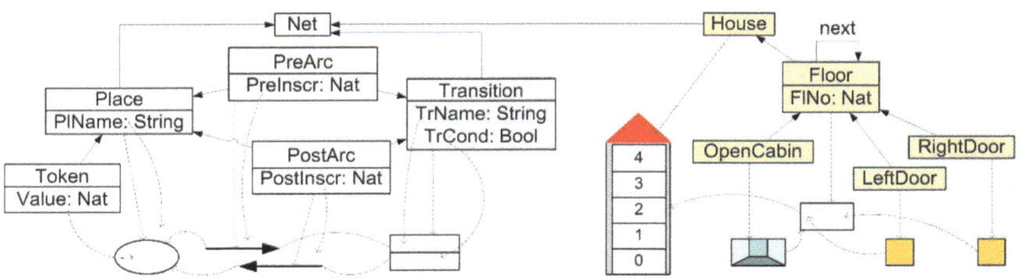

Fig. 6. Animation Alphabet for the Elevator Model

Based on the extension of the visual alphabet for AHL nets in Fig. 2 to the animation alphabet in Fig. 6, we have to extend now the simulation rules in Fig. 5 correspondingly. The resulting animation rules for the transitions start and up are presented in Fig. 7, while those for stop and down are given by the inverse rules for start and up, respectively. By applying first start and then up, the left snapshot of the animation view in Fig. 1 (b) is transformed into the right snapshot.

Moreover, by adding continuous animation operations to animation rules the resulting scenario animations are not only discrete-event steps but can

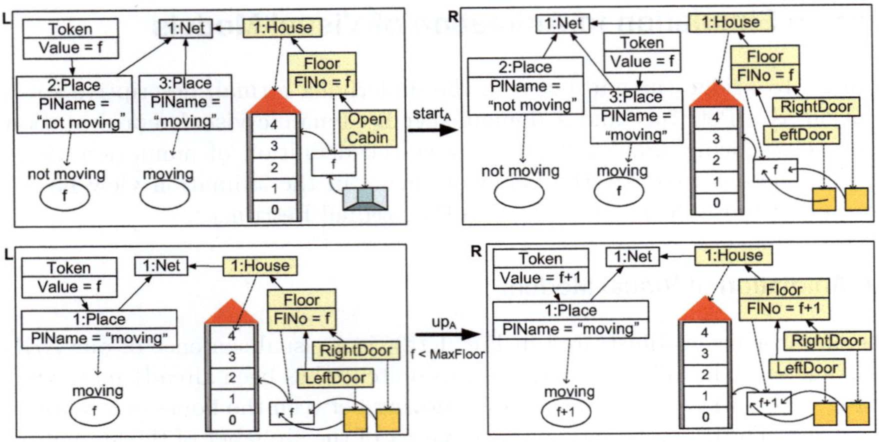

Fig. 7. Resulting Animation Rules for the Elevator Model

show the model behavior in a movie-like fashion. For the elevator, such animation operations are the opening and closing of the cabin doors (for the animation rules start and stop), and the up- and downward movements of the cabin doors (for the animation rules up and down).

4.2 General Construction and Correctness of Animation Rules

Now we present the main ideas for a general construction of animation rules from simulation rules as presented above. This construction is based on a model transformation from simulation to animation, short *S2A* model transformation. This *S2A* model transformation is again given by a typed attributed graph grammar. In general, *S2A* model transformation rules are non-deleting and are applied exactly once at each match to the graphs L and R of each simulation rule. *S2A* rules do not delete elements from the original simulation rules but only add elements from the domain specific extension of the animation alphabet.

For our elevator example, the *S2A* model transformation rules are given in Fig. 8. The tokens of the AHL net model the different locations of the elevator cabin. A token on place not moving corresponds to a cabin graphic with open doors, whereas a token on place moving is visualized by a graphic showing a cabin with closed doors. The position of the graphics relative to the elevator shaft depends on the value of the floor number. The *S2A* rules first add the basic elements for the animation view, i.e. the house symbol containing the elevator shaft. These elements are not changed during animation, and form the *animation background* (rule build house in Fig. 8).

After the generation of the animation background, depending on the marking of a place, an elevator cabin has to be drawn inside the floor corresponding to the token value on the place. For a marked place not moving, the symbol

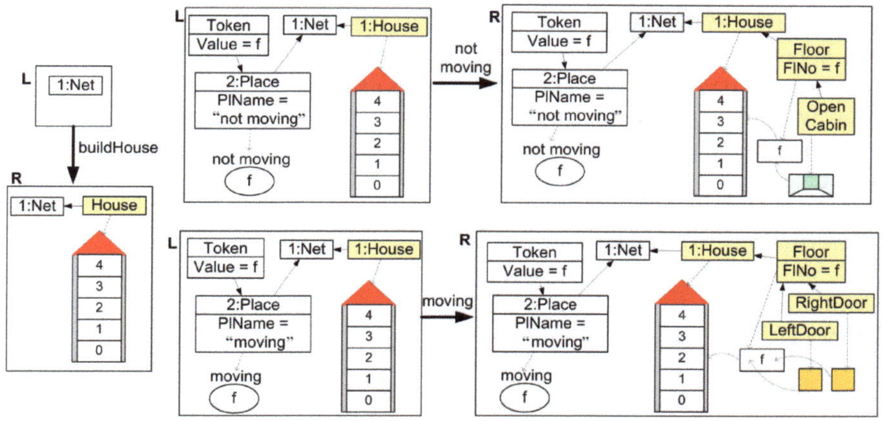

Fig. 8. S2A Rules for the Elevator Model

OpenCabin is generated. For a marked place moving, two doors of type LeftDoor and RightDoor are added (rules moving and not moving in Fig. 8).

In our paper [5], we give a general construction how to apply the *S2A* model transformation rules to simulation rules like those in Fig. 5, in order to obtain animation rules, like those in Fig. 7. Moreover, we present general criteria for the semantical correctness of animation w.r.t. simulation in the following sense, which are satisfied for our running example.

For each simulation step $S_1 \Longrightarrow S_2$ via a simulation rule sr with corresponding animation rule $ar = S2A(sr)$, there are animation models $A_1 = S2A(S_1)$ and $A_2 = S2A(S_2)$ and a corresponding animation step $A_1 \Longrightarrow A_2$ via ar, as depicted in the diagram to the right.

$$\begin{array}{ccc} S_1 & \xrightarrow{S2A} & A_1 \\ {\scriptstyle sr}\Downarrow & & \Downarrow{\scriptstyle ar} \\ S_2 & \xrightarrow{S2A} & A_2 \end{array}$$

5 Conclusion

In this paper we have given an overview of concepts for simulation and animation of visual models based on the formal theory of typed attributed graph transformation from [3]. This theory allows to show also formal correctness of visual modeling and simulation, as well as correctness of animation w.r.t. simulation (see [8, 5]).

The running example for simulation and animation of Petri nets has been modeled using the GenGED environment for visual language definition, simulation and animation based on graph transformation and graphical constraint solving [9, 1]. Animation operations are defined by coupling the animation rules to animations in SVG format (Scalable vector graphics), and can be viewed using suitable SVG viewers. In contrast to related approaches and tools, e.g. for the animation of Petri net behavior (like the SimPEP-tool for the animation of low-level nets in PEP [10]), the graph transformation framework offers a basis for a more general formalization of model behavior which is applicable to various Petri net classes and other visual modeling languages. Recently, the new tool environment TIGER [6, 12] has been developed at TU Berlin, supporting the generation of visual environments, based on the one hand on recent MDA development tools integrated in the development environment ECLIPSE [2], and on the other hand on typed attributed graph transformation to support syntax-directed editing and simulation.

References

1. R. Bardohl. A Visual Environment for Visual Languages. *Science of Computer Programming (SCP)*, 44(2):181–203, 2002.
2. Eclipse Consortium. *Eclipse – Version 2.1.3*, 2004. http://www.eclipse.org.
3. H. Ehrig, K. Ehrig, U. Prange, and G. Taentzer. *Fundamentals of Algebraic Graph Transformation.* EATCS Monographs in Theoretical Computer Science. Springer, 2006.
4. H. Ehrig, G. Engels, H.-J. Kreowski, and G. Rozenberg, editors. *Handbook of Graph Grammars and Computing by Graph Transformation, Volume 2: Applications, Languages and Tools.* World Scientific, 1999.
5. H. Ehrig and C. Ermel. From Simulation to Animation: Semantical Correctness of S2A Model and Rule Transformation. in preparation, 2006.
6. K. Ehrig, C. Ermel, S. Hänsgen, and G. Taentzer. Generation of Visual Editors as Eclipse Plug-ins. In *Proc. IEEE/ACM Intern. Conf. on Automated Software Engineering*, IEEE Computer Society, Long Beach, California, USA, 2005.
7. C. Ermel and R. Bardohl. Scenario Animation for Visual Behavior Models: A Generic Approach. *Software and System Modeling: Special Section on Graph Transformations and Visual Modeling Techniques*, 3(2):164–177, 2004.
8. C. Ermel, G. Taentzer, and R. Bardohl. Simulating Algebraic High-Level Nets by Parallel Attributed Graph Transformation. In H.-J. Kreowski, U. Montanari, F. Orejas, G. Rozenberg, and G. Taentzer, editors, *Formal Methods in Software and Systems Modeling*, vol. 3393 of *LNCS*. Springer, 2005.
9. GenGED Homepage. http://tfs.cs.tu-berlin.de/genged.
10. B. Grahlmann. The State of PEP. In *Proc. Algebraic Methodology and Software Technology*, vol. 1548 of *LNCS*. Springer, 1999.
11. H. J. Schneider. Graph Grammars as a Tool to Define the Behaviour of Process Systems: From Petri Nets to Linda. In *Proc. Graph Grammars and their Application to Computer Science*, vol. 1073 of *LNCS* pp. 7–12. Springer, 1994.
12. Tiger Project Team, Technical University of Berlin. *Tiger: Generating Visual Environments in Eclipse*, 2005. http://www.tfs.cs.tu-berlin.de/tigerprj.

Efficient Construction and Verification of Embedded Software

Sabine Glesner

Software Engineering for Embedded Systems
Technische Universität Berlin, Germany
glesner@cs.tu-berlin.de

Summary. Embedded software is rapidly growing in size and complexity while at the same time it is used in increasingly safety-critical applications. To cope with this situation, it is necessary to develop software in higher programming languages or even more abstract specification formalisms while at the same time providing methods to transform it into executable code whose semantics is provably the same as in the original specification or source program. In this paper, we present methods for the auto-omatic generation and verification of software transformations. These methods have not only been successfully applied in the area of optimizing compilers but can also be used in other areas as e.g. statechart transformations or hardware synthesis as well.

1 Introduction

Embedded software is one of the fastest increasing areas of information technology. Similarly to the exponential increase in hardware technology described by Moore's law, there is an analogous exponential growth in embedded soft-ware. For example, in the year 2003, an automobile had around 70 MB of embedded software, and one expects that this number increases to 1 GB until the year 2010. In general, software becomes a steadily increasing cost factor in embedded systems. This dramatic increase in the size as well as in the application ranges of embedded software implies that there is a steadily growing need for software engineering methods and tools to support the construction of embedded software. This need is even higher as the time to market needs to be as short as possible, while at the same time embedded systems need to be safe and robust, especially in safety-critical applications, as well as adaptive and easy to modify and maintain. While traditionally embedded software has been developed close to the hardware level, this does not keep pace with the aforementioned requirements. To fulfill these requirements, it is necessary to generate embedded software from higher programming languages or even more abstract specifications as e.g. statecharts while at the same time proving the correctness of these transformations.

Günter Hommel and Sheng Huanye (eds.), Embedded Systems Modeling – Technology, and Applications, 21–32.

In this paper, we introduce a method to generate and verify program and system transformations. Our approach concerning verification is twofold. First, we investigate if a certain transformation *algorithm* is correct, i.e. pre-serves the semantics of the transformed systems. Secondly, we are interested in the question if the transformation algorithm is also correctly *implemented.* From a practical point of view, this question is particularly interesting be-cause many translators are generated by generator tools and/or are too large to be verified entirely with today's verification tools. Our proofs are formulated within the interactive theorem prover Isabelle/HOL [NPW02]. Isabelle is a generic theorem prover that can be instantiated with different logics. The HOL (higher order logic) instantiation is particularly suited for dealing with the semantics of programming languages.

Our work contributes to the field of software engineering, especially soft-ware verification and compiler verification. Correct compilers are a necessary pre-requisite to ensure software correctness since nearly all software is written in higher programming languages and compiled into executable machine code afterwards. Our results show that important optimizations in realistic com-pilers can be formally verified. Moreover, our methods presented here can be applied to software transformations in general, as e.g. in software reengineer-ing, when using design patterns, in software composition and adaptation, in model driven architectures (MDA) etc.

This paper is organized as follows: In Section 2, we explain how program transformations can be verified by distinguishing between the correctness of the transformation algorithm and its implementation. Section 3 summarizes our results concerning the verification of transformation algorithms that transform intermediate program representations into machine code, while Section 4 shows how the implementation correctness of such transformations can be verified. In Section 5 we demonstrate that these verification methods can be applied in other areas as well, in particular when generating Java code from UML models. In Section 6, our long-term research vision is described. Section 7 discusses related work, and in Section 8 we conclude.

2 Correctness of Program Transformations

Concerning correctness of program transformations, one distinguishes between two different questions: First, one asks if a given transformation *algorithm* is correct, i.e. if it preserves the *semantics* of the transformed programs. Secondly, one investigates if its *implementation* is correct, i.e. if the compiler or tool executing the transformation algorithm is correctly implemented. The first notion of correctness is called *transformation correctness* while the second is denoted by *implementation correctness*. Implementation correctness has first been investigated in [Pol81, CM86]. In the following two subsections, we summarize previous work concerning these two notions of correctness.

2.1 Transformation Correctness

Typically, correctness proofs for program transformations consider refining transformations, i.e. transformations that do not alter the structure of programs but only specify in more detail how the individual computations are to be executed. For example, during the translation from higher programming languages into machine code, one needs to determine how complex data structures are mapped into the memory hierarchy of the processor. This implies that programs are transformed locally according to the *divide and conquer* principle and that the partial translations are put together subsequently. The correctness proofs follow this principle: correctness of translations is shown locally, and global correctness follows from them. Verifications according to this scheme can be found for example in [DvHG03].

2.2 Implementation Correctness

From a software engineering point of view, the question of implementation correctness is very important, especially if one takes into account that translators, for example compilers, typically are not manually implemented but generated from suitable specifications by generators. One could try to verify the generators themselves. Due to the sheer size of these systems, this option is not practical at all, in particular if one aims at machine proofs within theorem provers. With today's verification methodology, it is not (yet) possible to cope with systems of that scale. Instead one could try to verify the generated translators. This possibility is also not applicable, due to the same reasons.

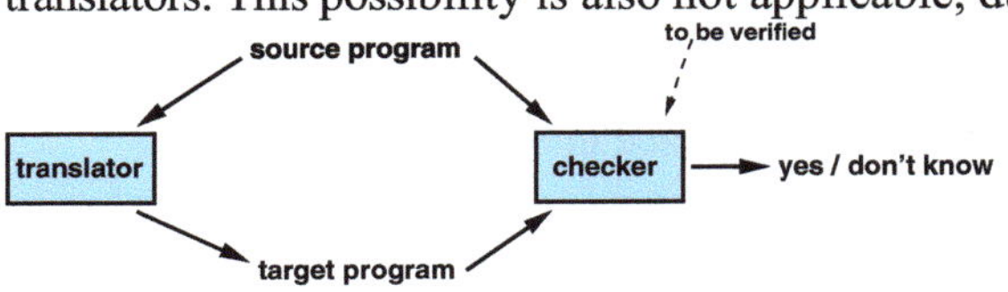

As a solution to this dilemma, *program checking,* also known as *translation validation* [PSS98], has been established as the method of choice [GGZ98]. Instead of verifying the translator, one only verifies its result. For this purpose, one enhances the translator by an independent checker, cf. Figure 1. For each run of the translator, the checker gets the input and output program of the translator as input and checks if they are semantically equivalent. If this is the case, then we have a proof that the transformation was indeed correct. In the negative case, we do not have anything. From a theoretical point of view, we cannot hope for more since program equivalence is undecidable in general. Nevertheless, from practical experience we know that this is an adequate and well-suited method. If one additionally verifies the checker with respect to a suitable specification, then, in case of a positive check, one gets a formally verified transformation result.

In compilers, this method for ensuring implementation correctness has proved its practicality in all frontend phases (lexical, syntactic and semantic analysis), especially because their transformation results are unique [HGG+99, Gle03b, GFJ04]. When checking for example the results of the syntactic analysis, one needs to make sure that the syntax tree computed by the compiler fits indeed to the context-free grammar of the programming language. This test can be computed in a top-down single traversal of the abstract syntax tree by checking for each node X_0 and its successor nodes $X_1, \ldots, X_n$ if there is a corresponding production $X_0 ::= X_1 \ldots X_n$ in the context-free grammar. Moreover, one needs to check that the syntax tree is a derivation for the source program by comparing the original source program text with the text derived by concatenating the leaves in the syntax tree to see if they are equal.

2.3 Open Problems in the Verification of Optimizing Compilers

In our work, we have investigated two hitherto unsolved problems: First we have dealt with the question how structure-modifying program transformation algorithms can be verified. Our results concerning this question are summarized in Section 3. As an example, we consider optimizing code generation from SSA (static single assignment) intermediate representations and discuss advantages and disadvantages of alternative proof approaches. In Section 4 we explain furthermore how the correctness of implementations of code generation can be ensured. Here as well as in other optimizing transformations one has to deal with the situation that many correct transformations exist and solutions are not unique. Applications of these proof methods in other areas such as the verification of Java code generation from UML models are summarized in Section 5.

3 Transformation Correctness for Code Generation

Machine code generation in compilers takes place after the source program has been transformed into an intermediate representation. We assume that the intermediate representation is in SSA (static single assignment) form [CFR+91], which expresses the essential data dependencies particularly clearly. In SSA form, every variable is assigned statically only once. SSA form can be obtained by suitable duplications and renamings of variables and is very well-suited for optimizations. As most intermediate representations, SSA form is based on basic blocks. Maximal sequences of operations that do not contain branches or jumps are grouped into basic blocks. The control flow of the program interconnects the individual basic blocks. Within basic blocks, computations are purely data-flow driven because an operation can be executed as soon as its operands are available. In this paper, due to space limitations, we only summarize our research concerning the verification of the data-flow driven parts of SSA computations in their translation into machine code, which is the semantically more interesting case. Details of these proofs can be found in [BGLM05, BG04].

Basic requirement for each formal verification of transformations is a formal semantics of the involved programming languages, in our case of SSA form. Since computations in SSA basic blocks are purely data-flow driven, one can think of them as directed acyclic graphs (DAGs). Note that basic block computations cannot be represented by terms in general because intermediate results may be used as input for several other operations, which gives rise to a graph structure. It is by far neither trivial nor unambiguous to formalize DAGs and graphs in general in a logical setting such as HOL. In our first try, we have formalized them as term graphs. In this approach, we have formalized DAGs in SSA form by representing them by an equivalent set of terms that contains shared subterms in duplicated form, cf. Figure 2. In the Isabelle/HOL theorem prover, we have specified them in the following representation, which is for better understanding only shown in a simplified form:

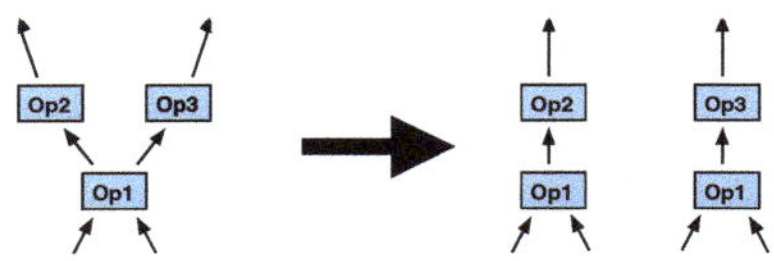

Fig. 2. Transformation of SSA DAGs into SSA Trees

datatype *SSATree = CONST value ident*
| *NODE operator SSATree SSATree value ident*

Without taking the specialties of Isabelle/HOL into account, one recognizes that this defines an inductive data type *SSATree* whose elements are either leaves or are assembled from subtrees. Moreover, we have defined a function *eval_tree* that takes an *SSATree* as input and evaluates it by specifying a result for all its nodes. The signature of this function is given by:

consts *eval_tree* :: *"SATree ⇒ SSATree"*

For the proof that machine code generation is semantics preserving, we need to define the semantics of the target machine as well as the mapping from SSA basic blocks into its machine code. For this purpose, we have chosen a relatively simple machine language whose operations correspond directly with the respective operations of the SSA basic blocks and whose programs are sequential lists of machine operations. For an exact formulation of this semantics, we refer to [BG04]. When mapping an SSA basic block to machine code, one has several possibilities, cf. Figure 3. For example, in the transformation of the code on the left, one must first compute op1 and has then the choice to compute op2 followed by op3 or op3 followed by op2. We could prove within Isabelle/HOL that each order is a valid code generation order that corresponds to a topological sorting of the nodes in the DAG.

Fig. 3. Transformation of SSA Form into Machine Code

With this proof, we have formally verified within the Isabelle/HOL theorem prover that each code generation order is correct that preserves the data dependencies. For this proof, we needed nearly 900 lines of proof code in Isabelle/HOL, which is rather a lot, especially if one takes into account that proofs in Isabelle/HOL must be done interactively by hand and cannot be generated automatically. At some parts, the proof was unnecessarily complicated because we had different induction principles (induction on SSA *trees* and induction on code *lists*). Due to these observations, we decided to set up a completely new simplified proof.

Based on the observation that machine code generation from SSA basic blocks is correct if the data dependencies are preserved, we have formalized basic blocks as partial orders on their operations. An operation o is smaller than another operation o' if o must be evaluated before o' because the result of o' depends directly or indirectly on the result of o. As in mathematical logic, we have formalized a partial order O as a set of tuples such that from $(x, y) \in O$ and $(y, z) \in O$ it follows that $(x, z) \in O$ (transitivity) and that from $(x, y) \in O$ and $(y, x) \in O$ it follows that $x = y$ holds (antisymmetry). Moreover, we have formalized code generation as a process that introduces further dependencies into the partial order. We could show that code generation is correct if the original SSA order is contained in the order implied by the machine code [BGL05].

Our experience with these two different proof alternatives in typical in the area of proof formalization within theorem provers. Both proofs are correct from a formal point of view as they both verify the desired correctness result. Nevertheless, the second alternative captures much better the essential proof ideas and is therefore the better alternative concerning comprehensibility and engineering of proofs. Formal (software) verification with theorem provers such as Isabelle/HOL is nowadays possible because of two main reasons: First, the performance of processors has increased so much that the search spaces of proofs can be evaluated sufficiently fast. Secondly, the user-friendliness of theorem provers has been improved significantly. Nevertheless, formal verification is still very expensive because of the necessary user interactions. These costs can be reduced if we manage to base proofs on general principles for system correctness (such as preservation of data flow dependencies), which can also be reused in other applications, cf. Section 5. Our results summarized in this section are an example for this.

4 Implementation Correctness for Code Generation

During code generation, the optimization variant of an **NP**-complete problem needs to be solved. Problems in **NP** are characterized by the fact that the correctness of solutions can be checked in polynomial time, whereas their computation is believed to need exponential time [Pap94]. This means that a non-deterministic Turing machine potentially needs to search an exponentially large computation space in which it will always find a solution in polynomial depth, cf. Figure 4. If one knows the path to such a solution (in complexity theory, it is called *certificate*), then one can compute the solution fast, i.e. in polynomial time. In most cases, the certificates of **NP**-complete problems are natural. For example, certificates for the SAT problem (satisfiability of propositional formulas) contain a satisfying assignment.

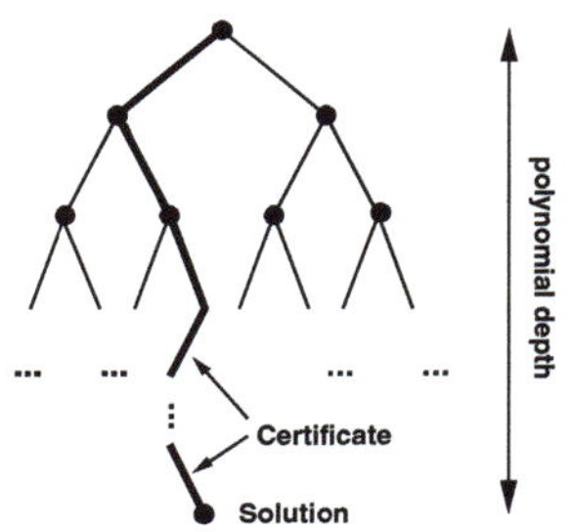

We use this property when ensuring the implementation correctness of optimizing transformations, cf. Figure 5. For this purpose, we modify the checking scenario used so far such that the translator outputs in addition to the target program also a certificate, which specifies how the target program has been computed. The checker gets this certificate as third input and uses it as follows: The checker computes again a target program from its input source program by using the certificate to find the target program efficiently. If this newly computed target program is identical with the one computed by the translator, then the checker outputs "yes", otherwise "don't know". We call this checking method *program checking with certificates* [Gle03a, Gle03b].

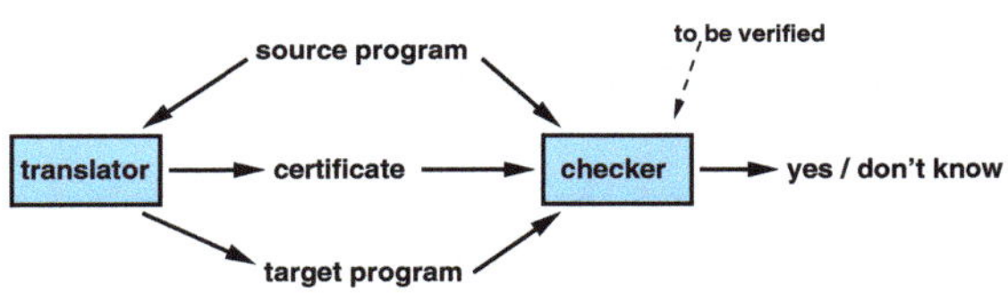

One may wonder what happens if the translator outputs a buggy certificate. If the checker manages to compute a correct result with a faulty certificate and if moreover this result is identical with the result computed by the translator, then the checker has managed to verify the result of the translator. In this process, it is irrelevant how the translator has determined its result, as long as the checker could reconstruct it with its (verified) implementation.

Let us look again at the roles that the translator (i.e. the code generator in our example here) and the checker are playing. The code generator behaves like a nondeterministic Turing machine. It searches for a solution and computes it. The checker acts as a deterministic Turing machine and only computes the solution, namely in the same way as the code generator by using the certificate. Hence, one could expect that the implementation of the checker is identical with some part of the implementation of the code generator, in fact with that part that computes the solution. Our experiments have confirmed this expectation. At first sight, this might seem unusual because originally one has used a checker to go for a test that is independent of the implementation to be checked. In our context here, the checking paradigm is modified. We use the checker here to separate the correctness-critical part of an implementation. The computation of the solution is correctness-critical. The search for a good solution is not correctness-critical because the quality of a solution does not influence its correctness.

	Code generator	Checker
lines of code in .h-Files	949	789
lines of code in .c-Files	20887	10572
total lines of code	21836	11361

The table on the left shows our experimental results. By using the method of program checking with certificates, we have designed and implemented a checker for a code generator, which has been implemented in the AJACS project (Applying Java to Automotive Control Systems) with industrial partners [GKC01]. The code generator has had more than 20.000 lines of code, the checker only about half of it. These numbers alone demonstrate that we could clearly reduce the verification cost. If one takes additionally into account that the search for a good or even optimal solution not contained in the checker is much more likely to be faulty and requires much more effort to be verified, then one clearly realizes that we have substantially reduced the verification costs by program checking with certificates.

5 Application in Related Areas

The verification and checking methods described in the previous two sections can be applied in other areas as well where source program or system descriptions are to be transformed into target programs or systems. In [BGL05], we have investigated how transformations of systems specified in the unified modeling language (UML) into Java code can be verified. In this research, we have formally verified within the Isabelle/HOL theorem prover that a certain algorithm for Java code generation from (simplified) UML specifications is semantically correct. This proof is part of more extensive ongoing work aiming to verify more general UML transformations and Java code generations.

6 Long-Term Research Vision

Many implementations of program and system transformations can be generated from suitable specifications. Examples are tools that generate executable code from statechart specifications. For example, the Statemate tool generates C or Ada code or VHDL or Verilog code from certain extended statechart specifications. In this case, the statechart specification is the source system and the C or Ada code or the VHDL or Verilog code are the target systems. Other examples are specifications of hardware systems, defined for example in VHDL or SystemC, which are now the source systems that are transformed into target hardware systems (e.g. FPGAs, ASICs, etc.). And last but not least there are the well-known compiler tools lex and yacc that take regular or LALR(1) context-free grammars, resp., as input and generate implementations that accept exactly the words that belong to the specified grammars. There exist further generator tools for basically all transformations taking place in all compilers phases. For the code generation phase in the backend of compilers, term rewriting mechanisms are widely used, cf. for example [NK97].

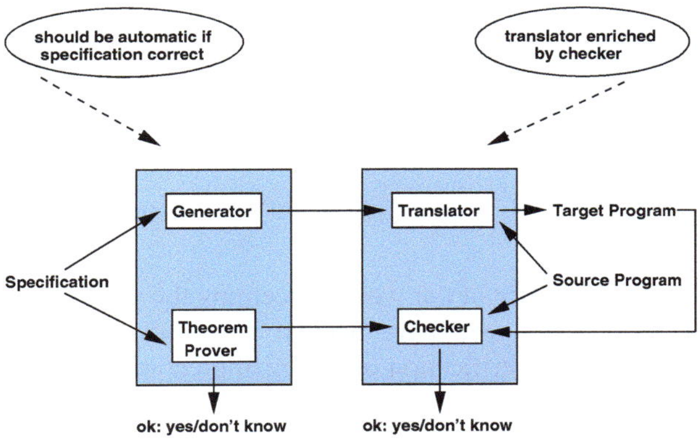

Fig. 6. Desired Overall System Architecture

Our work focuses on term-rewriting methods because they can be used generally for optimizing transformations of programs and systems. In the long run, we are aiming at a generator system that takes a specification as input that defines a transformation by means of term-rewriting mechanisms. Based on this specification, the generator shall not only output the corresponding translator but moreover a proof that the specified transformation is semantics-preserving as well as a checker that checks if the target programs and systems computed by the generated translator are correct, cf. Figure 6. A first experience with our tool (in the context of automatic construction of component systems) is reported in [GG06].

7 Related Work

The first work on the correctness of translation algorithms is [MP67] considering the correctness of a very simple compiler for arithmetic expressions. Early work on compiler verification with theorem provers is described in [Moo89]. Recent work has concentrated on transformations in compiler frontends. The formal verification of the translation from Java to Java byte code and formal byte code verification was investigated in [KN03]. The Verifix project [GGZ04] funded by the German Science Foundation (DFG) developed methods to construct correct compilers which are as efficient as typical commercial compilers. [DvHG03] describes Verifix results and considers the verification of a compiler for a Lisp subset in the theorem prover PVS. In [BGG05], the verification of a common compiler optimization, dead code elimination, is described.

Program checking was originally developed in the context of algebraic problems [BK95]. Since then, it has not only been applied successfully in the verification of compilers but also in other areas as well, for example in the verification of library routines [MN98] and in hardware verification [GD04].

The approach of proof-carrying code (PCC) [Nec97] is related to the verification methods for ensuring implementation correctness described in this paper. The principal idea of PCC is to require a code producer to generate a formal proof that the code meets certain safety requirements set by the code receiver. PCC is weaker than the verification methods considered here because it concentrates on necessary but not sufficient correctness criteria.

8 Conclusions

We have presented a summary of our work concerning the verification of program transformations. Moreover, we have shown how we have applied these verification methods in the verification of optimizing compilers. Furthermore, we have summarized how these methods can also be applied in other areas, for example in the generation of Java code from UML specifications. When verifying transformations, one has to deal with two different questions. First, one needs to investigate if a given transformation algorithm is correct, i.e. if it preserves the semantics of the transformed programs. Secondly, one needs to make sure that the implementation of the transformation algorithm is also correct. We have presented our solutions for both questions.

In our future work, we want to continue our work concerning the verifiation of transformations. In particular, we want to work on the completion of the generation and verification framework at which we are aiming, cf. Figure 6.

In this framework, translators shall be generated from suitable specifications such that the correctness proofs for the underlying transformation algorithms as well as checkers for their implementations are generated along the way. Moreover, we want to concentrate on transformations from system specifications (for example from UML models) to suitable programming languages or to machine code. And finally, we want to verify also non-functional properties such as memory consumption, time constraints, or security properties, which are especially important for embedded software.

Acknowledgement: Many thanks to the members and students of my research group who contributed to the results presented in this paper.

References

[BG04] Jan Olaf Blech and Sabine Glesner. A Formal Correctness Proof for Code Generation from SSA Form in Isabelle/HOL. In *Informatik 2004 – Informatik verbindet, 34. Jahrestagung der Gesellschaft für Informatik (GI), Band 2*, pages 449–558, 2004. Lecture Notes in Informatics (LNI).

[BGG05] Jan Olaf Blech, Lars Gesellensetter, and Sabine Glesner. Formal Verification of Dead Code Elimination in Isabelle/HOL. In *Proc. 3rd IEEE International Conference on Software Engineering and Formal Methods,* Koblenz, Germany, September 2005. IEEE Computer Society Press.

[BGL05] Jan Olaf Blech, Sabine Glesner, and Johannes Leitner. Formal Verification of Java Code Generation from UML Models. In *Proceedings of the 3rd International Fujaba Days 2005: MDD in Practice*. Technical Report, University of Paderborn, September 2005.

[BGLM05] Jan Olaf Blech, Sabine Glesner, Johannes Leitner, and Steffen Mülling. Optimizing Code Generation from SSA Form: A Comparison Between Two Formal Correctness Proofs in Isabelle/HOL. In *Proc. Workshop Compiler Optimization meets Compiler Verification (COCV 2005), 8th Europ. Conf. on Theory and Practice of Software (ETAPS 2005)*, pages 1–18, 2005. Elsevier, Electronic Notes in Theor. Comp. Sc. (ENTCS).

[BK95] Manuel Blum and Sampath Kannan. Designing Programs that Check Their Work. *Journal of the ACM*, 42(1):269–291, 1995.

[CFR+91] R. Cytron, J. Ferrante, B. K. Rosen, M. N. Wegman, and F. K. Zadeck. Efficiently Computing Static Single Assignment Form and the Control Dependence Graph. *ACM TOPLAS*, 13(4):451–490, October 1991.

[CM86] Laurian M. Chirica and David F. Martin. Toward Compiler Implementation Correctness Proofs. *ACM TOPLAS*, 8(2):185–214, April 1986.

[DvHG03] Axel Dold, Friedrich W. von Henke, and Wolfgang Goerigk. A Completely Verified Realistic Bootstrap Compiler. *International Journal of Foundations of Computer Science*, 14(4):659–680, 2003.

[GD04] Daniel Große and Rolf Drechsler. Checkers for SystemC Designs. In *Proc. Second ACM & IEEE International Conference on Formal Methods and Models for Codesign (MEMOCODE'2004)*, pages 171–178, 2004.

[GFJ04] Sabine Glesner, Simone Forster, and Matthias Jäger. A Program Result Checker for the Lexical Analysis of the GNU C Compiler. In *Proc. Workshop Compiler Optimization meets Compiler Verification (COCV 2004), 7th Europ. Conf. on Theory and Practice of Software (ETAPS 2004)*, pages 1–16, 2004. Elsevier, Electronic Notes in Theor. Comp. Science.

[GG06] Lars Gesellensetter and Sabine Glesner. Only the Best Can Make It: Optimal Component Selection. In *Workshop Formal Foundations of*

Embedded Software and Component-Based Software Architectures (FESCA 2006), 9th Europ. Conf. on Theory and Practice of Software (ETAPS 2006), 2006. Elsevier, Electr. Notes in Theor. Comp. Sc. (ENTCS).

[GGZ98] Wolfgang Goerigk, Thilo Gaul, and Wolf Zimmermann. Correct Programs without Proof? On Checker-Based Program Verification. In *Tool Support for System Specification and Verification, ATOOLS'98*, pages 108–123, 1998. Springer Series Advances in Computing Science.

[GGZ04] S. Glesner, G. Goos, W. Zimmermann. Verifix: Konstruktion und Architektur verifizierender Übersetzer (Verifix: Construction and Architecture of Verifying Compilers). *It–information Technology*, 46:265–276, 2004.

[GKC01] Thilo Gaul, Antonio Kung, and Jerome Charousset. AJACS: Applying Java to Automotive Control Systems. In *Conf. Proc. of Embedded Intelligence 2001, Nürnberg*, pages 425–434. Design & Elektronik, 2001.

[Gle03a] Sabine Glesner. Program Checking with Certificates: Separating Correctness-Critical Code. In *Proc. 12th Int'l FME Symposium (Formal Methods Europe)*, pages 758–777, 2003. Springer Verlag, LNCS Vol. 2805.

[Gle03b] Sabine Glesner. Using Program Checking to Ensure the Correctness of Compiler Implementations. *Journal of Universal Computer Science (J. UCS)*, 9(3):191–222, March 2003.

[HGG+99] A. Heberle, T. Gaul, W. Goerigk, G. Goos, and W. Zimmermann. Construction of Verified Compiler Front-Ends with Program-Checking. *Perspectives of System Informatics, PSI'99*, 1999. Springer LNCS Vol. 1755.

[KN03] Gerwin Klein and Tobias Nipkow. Verified Bytecode Verifiers. *Theoretical Computer Science*, 298:583–626, 2003.

[MN98] Kurt Mehlhorn and Stefan Näher. From Algorithms to Working Programs: On the Use of Program Checking in LEDA. *23rd Int'l Symp. on Math. Found. of Comp. Sc. (MFCS'98)*, 1998. Springer LNCS Vol. 1450.

[Moo89] J. S. Moore. A Mechanically Verified Language Implementation. *Journal of Automated Reasoning*, 5(4):461–492, 1989.

[MP67] John McCarthy and J. Painter. Correctness of a Compiler for Arithmetic Expressions. In *Mathem. Aspects of Computer Science, Proc. Symposia in Applied Mathematics*, pages 33–41. American Mathem. Society, 1967.

[Nec97] George C. Necula. Proof-Carrying Code. In *Proc. 24th ACM SIGPLAN-SIGACT Symp. on Principles of Progr. Languages (POPL'97)*, 1997.

[NK97] A. Nymeyer and J.-P. Katoen. Code generation based on formal BURS theory and heuristic search. *Acta Informatica 34*, pages 597–635, 1997.

[NPW02] Tobias Nipkow, Lawrence C. Paulson, Markus Wenzel. *Isabelle/HOL: A Proof Assistant for Higher-Order Logic*. Springer LNCS Vol. 2283, 2002.

[Pap94] C. H. Papadimitriou. *Computational Complexity*. Addison-Wesley, 1994.

[Pol81] Wolfgang Polak. *Compiler Specification and Verification*. Springer Verlag, Lecture Notes in Computer Science, Vol. 124, 1981.

[PSS98] A. Pnueli, M. Siegel, E. Singerman. Translation validation. In *Tools and Algor. for the Constr. and Anal. of Systems*, 1998. Springer LNCS 1384.

Embedded Systems Design Using Optimistic Distributed Simulation of Colored Petri Nets

Michael Knoke, Dawid Rasinski, and Günter Hommel

Real-Time Systems and Robotics
Technische Universität Berlin, Germany
{knoke,ronin,hommel}@cs.tu-berlin.de

1 Introduction

Extensive simulations are required before embedded systems for mass production can be implemented and deployed to the market. Especially performance and reliability evaluations accompany the architectural design. *Colored* Petri nets (CPN) have proven to be appropriate for this task. The high complexity of embedded systems requires increased performance simulating the behavior of a dedicated system using distributed simulation.

There has been significant work in the area of distributed simulation of Petri nets (PN) in the past few years, but properties of CPNs which cause immediate state changes, are either not allowed or limit the distribution of corresponding model parts, so that they must be processed sequentially. A new logical clock concept that fully characterizes causality in CPNs has been proposed by Knoke et al. [1,2]. It guarantees a true identification of simultaneous concurrent events and correct ordering based on their past priority conflicts.

This logical clock enables a fine-grained logical model partitioning. It has been shown in [2] that a more fine-grained partitioning optimizes rollback execution, simplifies logical clock implementation, and facilitates a distribution of model parts to logical processes (LP). Each partition has its own virtual time and is independent, so that the mapping allows almost arbitrary associations of model parts to computing nodes. Dynamic remapping at runtime benefits from more accurate information about the virtual time progress of model components.

As a consequence, traditional model partitioning algorithms for distributed Petri net simulation (DPNS) cannot be used or at least do not exploit the full properties of the new logical clock concept. Hence, the main motivation for our work is to introduce dynamic remapping approaches which take advantage of the fine-grained logical partitioning. It is often impossible to obtain a good partitioning from the model structure because of CPN features which depend on global dependencies such as global guards, place capacities andglobal result measures. An efficient optimistic simulation may require widely dispersed

Günter Hommel and Sheng Huanye (eds.), Embedded Systems Modeling – Technology, and Applications, 33–42.

model parts with global dependencies to be simulated by the same LP, as shown in [3]. Our approach allows to achieve a good balance with simple heuristics starting from an arbitrary initial mapping to computing nodes. Some substantial performance measures of the new approaches will be shown in this paper.

The paper first presents the proposed fine-grained partitioning in Sect. 2. The subsequent Sect. 3 introduces new remapping approaches which ease the migration decision based on the virtual time differences of independently simulated model parts. A discussion and comparison of these approaches are presented in Sect. 4. Finally, concluding remarks are given in Sect. 5.

2 Previous Work

Distributed simulation of colored Petri nets is difficult due to their special properties. Some of our previous work regarding DPNS having an impact on dynamic load-balancing is presented in this chapter. First, we present the fine-grained logical partitioning which is a basis for our dynamic remapping approach.

2.1 Fine-grained Logical Partitioning

The simulation is composed of N sequential event driven LPs that do not share memory and operate asynchronously in parallel. Unlike in other optimistic DPNS algorithms (e.g. introduced in [4]), an *Atomic Unit* (AU) is defined to be the smallest indivisible part of the model

$$AU = \left\{au_1,\ au_{2,\ldots},au_{|AU|}\right\}.$$

AUs are distributed over the available LPs, which we denote by the node-distribution function *Node*.

$$Node: AU \rightarrow LP$$

The region r^{AU} of an AU is defined as the minimal set of transitions T_r which have common input places $P^{in}(T_r)$. These input places and corresponding arcs $A(T_r)$ are joined as well. Formally we denote:

$$r^{AU} = (P_r \subseteq P, T_r \subseteq T, A_r \subseteq A)$$

Let $T^{out}(p)$ the set of output transitions of place p, then:

$$\forall t \in T\colon P_r(t) = \{p_a | p_a \in P^{in}(t) \vee \exists p_b \in P_r,\ p_a \neq p_b :$$
$$T^{out}(p_a) \cap T^{out}(p_b) \neq \emptyset\}$$
$$T_r(t) = \bigcup_{p \in P_r} T^{out}(p)$$

At least one AU is assigned to each LP_i which is running as a process on one physical node N_i. A communication interface attached to the LPs is responsible for the propagation of messages to remote LPs and to dispatch incoming messages to local AU s. AU s on the same LP communicate directly to avoid additional message overhead. Each LP_i and AU_j has access only to a partitioned subset of the state variables $S_{P,i} \subset S$ and $S_{U,j} \subset S_{P,i}$, disjoint to state variables assigned to other LP s or AU s. State variables of LP_i are the set of state variables of all local AU s $S_{P,i} = \cup S_{U,j}(\forall j)$. The simulation of local AU s is scheduled within each LP in a way that avoids local rollbacks [1].

This AU approach reduces memory and computational costs of the logical clock. Also, a rollback of complete LPs will not happen. Other advantages as well as performance measures can be found in [2]. Each AU has its own virtual simulation time and stores its state for each local event independently from other AUs. This characteristic is essential for a migration to other LPs at runtime. As a result, dynamic load balancing is a remapping of statically composed logical partitions (AUs) and unlike other load balancing algorithms does not require expensive repartitioning.

3 Remapping of Model Parts

Load management research has a long history but is mostly based on assumptions that do not apply to DPNS. Dynamic load balancing tries to equalize some metric, typically utilization, across all nodes, usually by migrating existing objects. Many algorithms have been proposed for moving load from one node to another, given that dynamic load management is being used. While a lot of these methods offer suggestions on how to perform load management, only a few solve the problem of managing load in a virtual time-based system. Optimistic systems need to balance the amount of useful work done, rather than balancing the total work.

The CPN model in Fig. 1 shows exemplarily that intuitive partitioning is not appropriate to find an optimal distribution. This model consists of three completely independent parts, each logically partitioned into two AUs. A simulation on three LPs would suggest a mapping where each part is simulated on one LP. The same distribution would arise for communication costs as a primary criterion for migration. But in reality this partitioning slows down the simulation by 12 percent compared to a sequential simulation on one processor. The optimal partitioning gains 10 percent speed-up on three LPs for this model. So we conclude that it is not sufficient to simply analyze structural properties of CPNs for good performance results.

Important results were given by Reiher and Jefferson [5] presenting a virtual time based dynamic load management in the Time-Warp-OS. Reiher and Jefferson argued that virtual time is not unambiguous for repeated simulation runs, so that it should not be used as a criterion for load balancing. This is not true for our simulation framework because the logical clock explained in [1] ensures accurate occurrence of events.

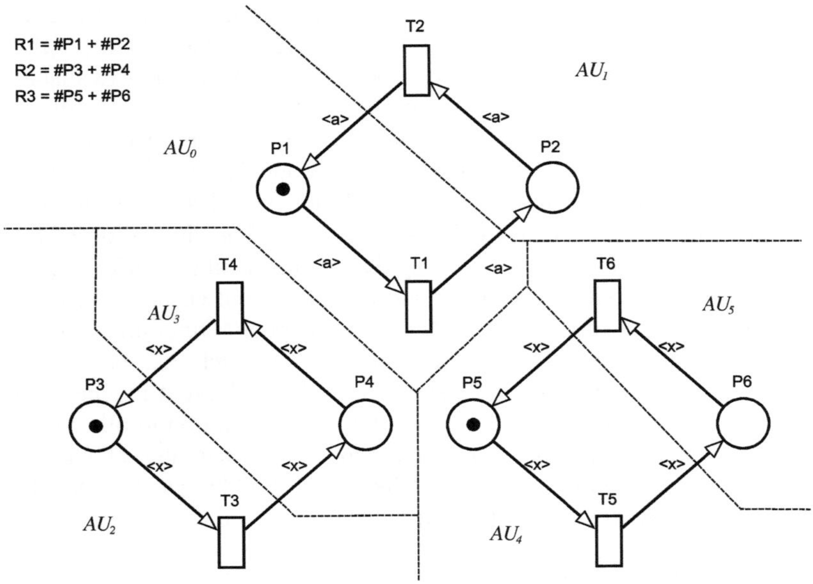

Fig. 1. Simple CPN model consisting of three independent parts is important to calculate the difference quotient to relate the number of time steps in a certain interval and the length of that interval.

Virtual Time Progress (VTP) is an inverse measure for load. It reflects, how fast a simulation process continues in virtual time. The ideal case of an undisturbed system should lead to an almost identical VTP on all nodes. A system disturbance results in a time displacement between nodes. Various load balancing approaches try to measure the causes, as done by Reiher and Jefferson [5] using the effective load, or by Nicol and Mao [6] using the work load of a LP. But it is difficult to capture all possible causes and unconsidered causes that prevent an optimal load balancing decision. Our approach concentrates on the symptoms, namely the virtual time progress of each node, which indirectly captures all causes. We try to determine how far this is sufficient without additional measures.

3.1 Load Balancing Decisions

In this paper we concentrate on measuring the virtual time progress of LPs and AUs because it is independent from system properties and in contrast to traditional DPNS such as Time-Warp, easy to apply. Additionally, it takes the biggest advantage of fine-grained partitioning. On closer examination it is obvious that the progress of virtual time is in principle the first derivative of virtual time to real time. But virtual time has a discrete character, so it is important to calculate the difference quotient to relate the number of time steps in a certain interval and the length of that interval.

Virtual Time Progress of LPs

The virtual time progress of LPs can be seen as a measure for workload. Applied to our AU approach, VTP_{LP} is the minimum virtual time of all AUs on LP. If $VTP_{LP1} < VTP_{LP2}$ then LP_1 needs more time to simulate its associated model part. Furthermore LP_1 may cause rollbacks of LP_2. A possible situation is illustrated in Fig. 2 where LP_1 proceeds quickly but is rolled back continuously by LP_0.

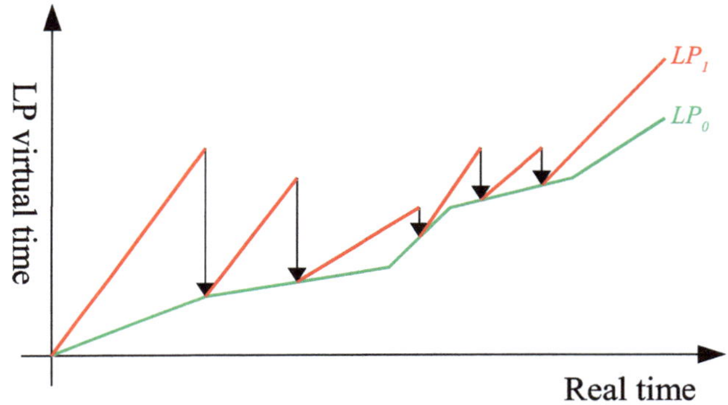

Fig. 2. Virtual time difference of LPs

The most heavily loaded LP is the LP with the lowest VTP value and the least heavily loaded LP the one with the highest VTP value. A bottom line of this consideration leads to a load balancing decision where the AU with the lowest virtual time of the most heavily loaded LP should be migrated to the least heavily loaded LP. An estimation of the migration profit is directly observable by the VTP differences. First, we introduce the average firing time $\overline{T}_{au}$ which is the average time distance of created events on $au \in AU$. Likewise,

$$\overline{T}_{\max,lp} = \max_{\forall au \in LP}(\overline{T}_{au})$$

is the maximum average time distance of all events created on $lp \in LP$. Migration profit is ensured for two LPs n and m if $VTP_n - VTP_m > \Delta t_{nm,\,max}$ whereas $\Delta t_{nm,\,max} = \tau_{LP} * max(\overline{T}_{max,n}, \overline{T}_{max,\,m})$. Factor τ_{LP} ($\tau_{LP} \geq 1$) is an additional time difference tolerance. If this algorithm has determined two LPs for migration, the sender will locally select the AU with the minimum virtual time.

Virtual Time Progress of AUs

Another possible criterion for migration is a direct comparison of the virtual time progress of all AUs. In contrast to the previous approach it considers the time difference between all AUs and hence allows for a more precise decision. A small time difference $\Delta t_{n,max}$ between the fastest and slowest AU identifies an ideal source LP n for migration and is calculated as follows: $\Delta t_{n,max} = \tau_{AU} * \overline{T}_{max,n}$ whereas τ_{AU} is once again a time difference tolerance value. Experiments have shown major differences of these time progress values. This should provide an indication that a good tolerance value for τ_{AU} will be difficult to obtain.

Connections between AUs

The number of connections between AUs can be seen indirectly as a simplified measure for communication costs, assuming that a higher number of connections leads to more communication. The intention of this criterion is to collect multi-connected AUs on the same LP. But it is based on measures captured once and therefore not applicable as a single criterion because it has no relation to the current workload of the simulation. So it is used in conjunction with one of the other approaches. Assuming source LP n and destination LP m for migration are already identified. We try to find AU_j on LP n which has the maximum number of connections c_{mj} to LP m, whereas ω_{kj} is the number of connections from AU_k to AU_j. As a constraint to ensure profit of this migration, we require $c_{nj} < c_{mj}$.

$$c_{mj} = \sum_{\forall k: AU_k \in LP_m} (\omega_{kj} + \omega_{jk})$$

Communication between AUs

Measuring real communication costs between AUs could improve the accuracy of a migration decision, particularly with regard to the previous approach that just considers the number of connections. Nevertheless, this criterion is not suitable as a single criterion because we don't have message communication among AUs within the same LP, so corresponding costs cannot be measured. But it offers valuable hints for the first two approaches. The calculation is similar to the previous approach, but tries to find AU_j on LP n which has maximal communication κ_{mj} to LP m.

$$\kappa_{mj} = \sum_{\forall i: AU_i \in LP_m} (\kappa_{ij} + \kappa_{ji})$$

4 Performance Tests

In the course of our study we have designed numerous models to apply an efficient load balancing algorithm. All experiments have been conducted on a 16 node, dual Intel Xeon 1.7 *GHz* processor, Linux cluster with 1 GB memory on each node. SCI has been used as a high-speed network connected as a 2D-Torus.

Only one processor on each cluster node has been used for all experiments to simplify comparison of the results. In contrast to Time-Warp, the AU approach allows a truly distributed and efficient simulation of even simple models to exploit parallelism. To discuss the performance of the proposed remapping schemes, we abandon a demonstration of best performing models. It is obvious that very complex models with thousands of places and transitions and with less synchronization between model parts will achieve the highest speed-up in a distributed environment. But because of their complexity these models are not qualified to be discussed here. Otherwise, a distributed simulation of simple models on fast processors will never achieve good overall performance. However, simple models are more easily understandable, so they are used in this paper for comparison.

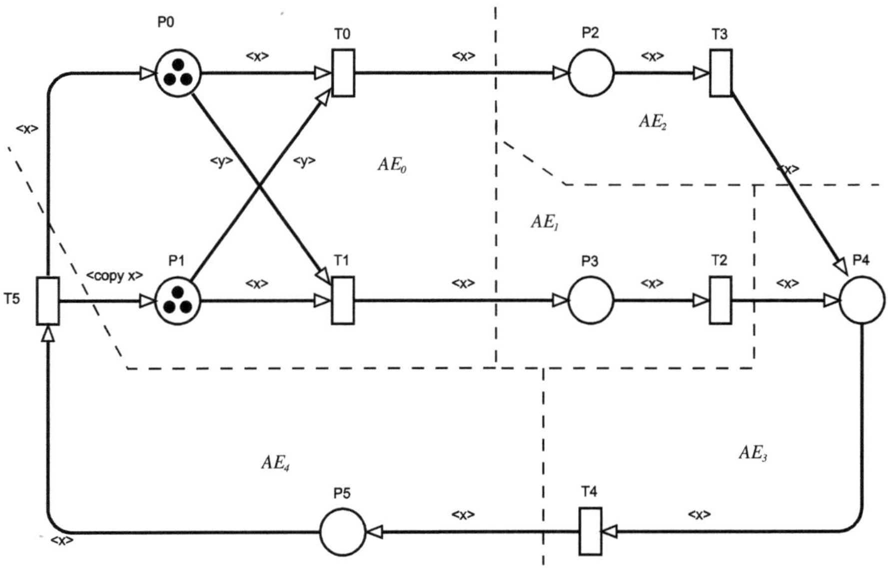

Fig. 3. CPN model with conflicts and concurrency

Figure 1 shows a very simple CPN model, already discussed in Sect 3. Regardless of the simplicity, the three independent model parts impose high requirements on the load balancing algorithm because we have seen that the most intuitive distribution performs poorly. A second model is illustrated in Fig. 3. It is partitioned into 5 AUs and has a low potential of parallelism. In most instances, the behavior of AUs depends on one or more other AUs. This may restrict parallel execution and leads to activities on a single AU or frequent rollbacks. A high speed-up of a parallel simulation cannot be expected. That is why this model is very sensitive to distribution decisions on less than 5 nodes and poses a challenge for load balancing algorithms.

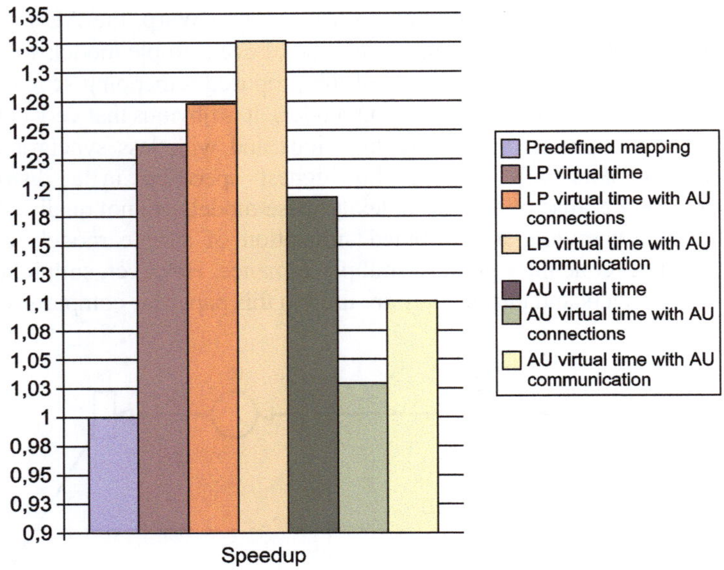

Fig. 4. Results of the different remapping strategies for model Simple Triplet

Each model is statically partitioned into a set of AUs, as described in Sect. 2.1. Also, a predefined (non-optimal) initial mapping of AUs to LPs has been performed in all experiments. The remapping approaches discussed in Sect. 3 have been applied for a distributed simulation of both models on three cluster nodes. Figure 4 presents the speed-up which could be obtained for the simple CPN model. It shows that the virtual time progress of LPs allows very good dynamic remapping particularly with analysis of communication costs as secondary criterion. A speed-up of 1.3 compared to the initial partitioning is excellent in consideration of the unfavorable general framework: simple model and high speed cluster. The same conclusion arises from the second CPN model even though the impact of the secondary criterion is different. Results of this experiment are depicted in Fig. 5.

The virtual time progress of AUs promised better results. We traced this problem back to the tolerance values. Either a stable final state could not be reached and frequent remapping increased the computational effort, or too few migrations took place. The tolerance design is problematic as discussed earlier. A dynamic remapping based on the virtual time analysis on LP level also depends on the corresponding tolerance value, but not as much as the AU approach. This is shown in Fig. 6 for different remapping strategies.

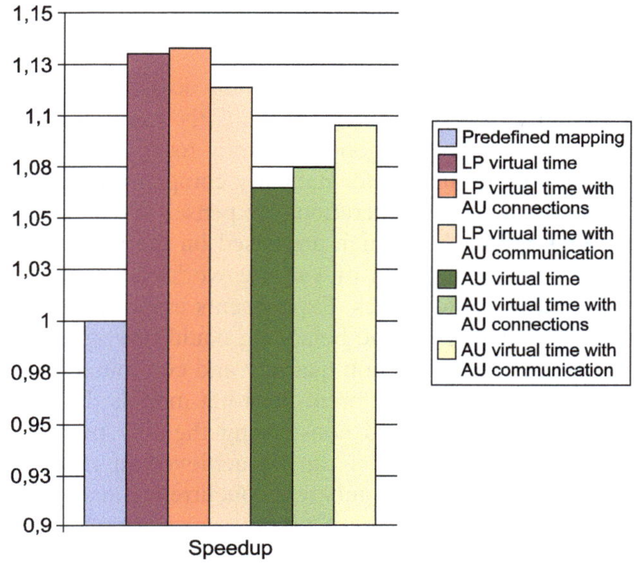

Fig. 5. Results of the different remapping strategies for model Complex

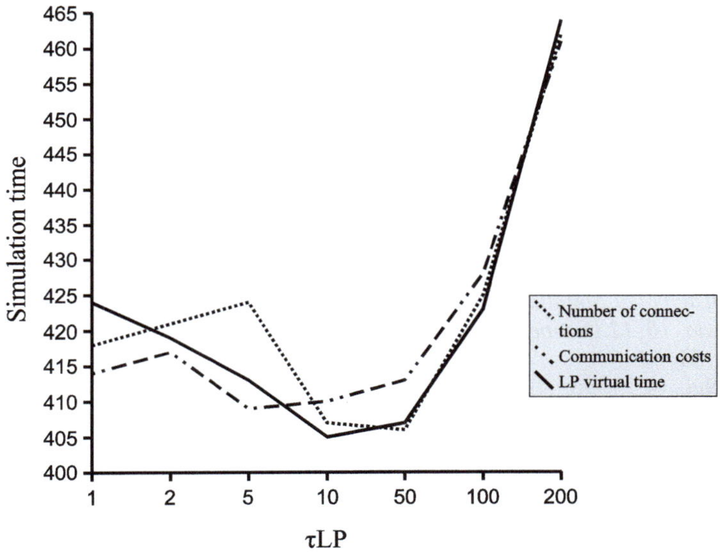

Fig. 6. Desired Overall System Architecture

5 Conclusion

Today's requirements for flexibility and maximum scalability for a simulation of embedded systems necessitate the use of CPN models whose efficient execution requires a preferably good dynamic load balancing. By using finegrained partitioning, partitions are statically composed and just need to be remapped at runtime with low operational expense. We introduced several dynamic remapping approaches which are based on the virtual time progress of LPs and AUs. Thus, symptoms of varying workload could be measured without capturing all possible causes. Experiments emphasized the feasibility of this approach. The quality of load balancing could be slightly enhanced by secondary criteria, such as connection quantity and communication costs. The speed-up has been shown using different Petri net models. In the majority of cases the speed-up was very good considering the low model complexity. Dramatically better speed-up results can be achieved in real-world models with high complexity exploiting the inherent concurrency in such applications.

References

1. Knoke, M., Kuehling, F., Zimmermann, A., Hommel, G. 2004, "Towards Correct Distributed Simulation of High-Level Petri Nets with Fine-Grained Partitioning." In J. Cao, L. T. Yang, M. Guo, et al., eds., *Proceedings of the Second International Symposium: Parallel and Distributed Processing and Applications (ISPA 2004), December 13–15, 2004, Hong Kong, China*, Lecture Notes in Computer Science 3358, IEEE, ACM, Springer-Verlag, Berlin, 64–74.
2. Knoke, M., Kuehling, F., Zimmermann, A., Hommel, G. 2005, "Performance of a Distributed Simulation of Timed Colored Petri Nets with Fine-Grained Partitioning." In D. Tutsch, ed., *Proceedings of the 2005 Design, Analysis, and Simulation of Distributed Systems Symposium (DASD 2005), April 3–7, 2005, San Diego, California, USA*, SCS, 63–71.
3. Knoke, M., Hommel, G. 2005, "Dealing with Global Guards in a Distributed Simulation of Colored Petri Nets." In A. Boukerche, S. J. Turner, D. Roberts, G. K. Theodoropoulos, eds., *Proceedings of the 9th IEEE International Symposium on Distributed Simulation and Real Time Applications, DS-RT 2005, Montreal, Canada, 10–12 October 2005*, IEEE, IEEE Computer Society, Los Alamitos, CA, USA, 51–58.
4. Ferscha, A. 1996, "Parallel and Distributed Simulation of Discrete Event Systems." In A. Y. Zomaya, ed., *Parallel and Distributed Computing Handbook*, McGraw-Hill Inc., 1003–1041.
5. Reiher, P. L., Jefferson, D. R. 1990, "Virtual Time Based Dynamic Load Management in the Time Warp Operating System." In *Proceedings of the SCS Multiconference on Distributed Simulation*, vol. 22, 103–111.
6. Nicol, D. M., Mao, W. 1995, "Automated Parallelization of Timed Petri-Net Simulations." Journal of Parallel and Distributed Computing, 29, no. 1: 60–74.

Extended Reward Measures in the Simulation of Embedded Systems With Rare Events

Armin Zimmermann

Real-Time Systems and Robotics
Technische Universität Berlin, Germany
azi@cs.tu-berlin.de

1 Introduction

Model-based evaluation of embedded systems is an important aid in their design phase. This is especially true for non-functional properties which are of less importance in other areas. Examples are real-time capabilities, fault-tolerance, and performance while taking into consideration failures as well as heavy load situations.

Performance evaluation requires a model with performance measures and an evaluation technique implemented in some software tool. In the following we adopt stochastic discrete-event systems (SDES) as a family of models such as queueing networks or Petri nets. They have proven to be applicable to a wide area of technical systems in the past. Numerical analysis or simulation can be used for the derivation of performance measures. Exact analysis techniques are however restricted by the size of the reachability graph that can be handled as well as by not allowing parallel activities with non-Markovian behavior. We study simulation in the following, which is not subject to these restrictions.

There are many examples of safety-critical embedded systems in which the designer is interested in the probability of reaching a state of potential catastrophic behavior. Very low probabilities of reaching a state of interest requires a vast amount of events to be generated. This case is called *rare event simulation* due to the low ratio of significant samples w.r.t. the overall event number. There are numerous applications in which probabilities in the range of 10^{-8} or less need to be quantified. Standard simulation methods are not applicable in practice because they would require prohibitively long run times before achieving statistical accuracy. Variance reduction techniques such as control variables, common random numbers and others do not solve this problem, because it may easily happen that no significant event is ever generated in an acceptable time.

Several approaches have been investigated in the literature to overcome this problem; overviews are e.g. given in [6, 4, 2]. They have the common

Günter Hommel and Sheng Huanye (eds.), Embedded Systems Modeling – Technology, and Applications, 43–52.

goal to make the rare event happen more frequently in order to gain more significant samples out of the same number of generated events.

Importance sampling [3] changes the model in a way that lets the rare event happen more often. Its main drawback is that it usually requires deep insight into the model, and is thus considered for simple models only [4].

The second main technique for rare-event simulation is *importance splitting*, which has been introduced in the context of particle physics and later used in simulation. Its underlying idea to generate more samples of the rare event is to follow paths in the simulated behavior that are more likely to lead to an occurrence of the event. It thus requires an algorithm to decide which paths to take and which to discard; the subsequent section briefly shows how this is usually done.

There are two different approaches in the literature, one under the name of *splitting* [2] and the other called *RESTART* [15,13] (which is short for *repetitive simulation trials after reaching thresholds*). Splitting is applied to estimate the (small) probability of reaching a set of states out of an initial state before the initial state is hit again. It requires the system to return to this state infinitely many times. RESTART is considered to be less restrictive than splitting w.r.t. the types of performance measures to be computed. Moreover, measures can be estimated both in transient and steady-state. It has been extended significantly in [15] by the authors that introduced this version of splitting techniques. The application to arbitrary models (e.g. in a software tool or a model framework such as SDES) is easier because it can use the model as a black box, as long as the measure of interest is defined as described below.

Recent treatments of the RESTART technique can be found in [13,14]. Surveys of the more general field of rare-event simulation include [6,2,4]. Stochastic Petri net applications of RESTART have been implemented in the TimeNET software tool [8,7] as well as in SPNP [11]. Other implementations of RESTART include ASTRO [12]. Importance sampling techniques are e.g. included in SAVE [5] and UltraSAN [9], the latter for stochastic activity networks. Different ways of implementation and their characteristics are compared in [1].

We consider the RESTART method out of the mentioned reasons here. Its standard usage for the estimation of a low probability is extended in Section 3 to allow arbitrary reward measure types. Section 3.1 briefly covers reward measures, and Section 3.2 describes the proposed algorithm.

2 The RESTART Method

Assume that the goal of a simulation is to estimate the probability $\mathrm{P}\{A\}$ of being in a set of state A in steady state, and that significant samples are generated only rarely due to the model. Let the set of all reachable states of a model be denoted by B_0, and the initial state of the system by σ_0. This situation is sketched in the Venn diagram in Figure 1.

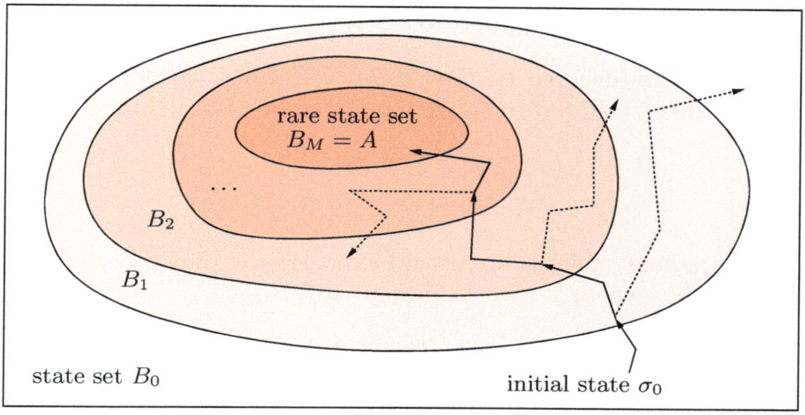

Fig. 1. State sets and paths in a rare-event simulation with splitting

A standard simulation would require a very long run time until A has been visited sufficiently often to estimate $P\{A\}$. The idea of importance splitting techniques in general is to let A be visited more frequently by concentrating on promising paths in the state set. The line in Fig. 1 from the initial state σ_0 into A depicts one possible "successful" path. A standard simulation would usually deviate from this and take paths leading "away" from A. If we can find a measure of "how far away" from A a state is, it becomes possible to decide which paths are more likely to succeed and should be followed more frequently. Such a measure can often be deduced from the actual application: rare failures may be the result of a continuous wear-and-tear, and unavailability or losses due to blocking happen after a buildup in buffers. Corresponding regions of the state set with increasing probability of reaching A can then be defined. The samples resulting from such a guided simulation must of course be corrected by the ratio w.r.t. the probability that the path would have been taken in a standard simulation.

Formally, define M subsets $B_1 \dots B_M$ of the overall state space B_0 such that

$$A = B_M \quad \text{and} \quad B_M \subset B_{M-1} \subset \dots \subset B_1 \subset B_0$$

The conditional probabilities $P\{B_{i+1} \mid B_i\}$ of being in an enclosed set B_{i+1} under the precondition of being in B_i are much easier to estimate than $P\{A\}$, because every one of them is not rare if the B_i are chosen properly. The measure of interest can then be obtained from the product of the conditionals (obviously $P\{B_0\} = 1$).

$$P\{A\} = \prod_{i=0}^{M-1} P\{B_{i+1} \mid B_i\}$$

States visited during a simulation must be mapped to the respective sets B_i. An *importance function* f_I returns a real value for each state $\sigma \in B_0$.

$$f_I : B_0 \to \mathbb{R}$$

A set of *thresholds* (denoted by $Thr_i \in \mathbb{R}, i = 1 \dots M$) divides the range of importance values such that the state set B_i can be obtained for a state[1].

$$\begin{aligned} \forall i \in \{0 \dots M\} \quad &: \quad Thr_{i+1} > Thr_i \\ \sigma \in B_i \quad &\iff \quad f_I(\sigma) \geq Thr_i \end{aligned}$$

Setting of threshold values is discussed below. We say that the simulation is in a *level* i if the current state σ belongs to $B_i \setminus B_{i+1}$.

$$Level(\sigma) = i \iff Thr_i \leq f_I(\sigma) < Thr_{i+1} \tag{1}$$

An importance splitting simulation measures the conditional probability of reaching a state out of set B_{i+1} after starting in B_i by a Bernoulli trial. If B_{i+1} is hit, the entering state is stored and the simulation trial is split into R_{i+1} trials. The simulation follows each of the trials to see whether B_{i+2} is hit and so on. A trial starting at B_i is canceled after leaving B_i if it did not hit B_{i+1}. Simulation of paths inside B_0 and $B_M = A$ is not changed.

An estimator of $\mathrm{P}\{A\}$ using R_0 independent replications is then [11]

$$\widehat{\mathrm{P}\{A\}} = \frac{1}{R_0 R_1 \dots R_{M-1}} \sum_{i_0=1}^{R_0} \dots \sum_{i_{M-1}=1}^{R_{M-1}} \mathbf{1}_{i_0} \mathbf{1}_{i_0 i_1} \dots \mathbf{1}_{i_0 i_1 \dots i_{M-1}}$$

if we denote by $\mathbf{1}_{i_0 i_1 \dots i_j}$ the result of the Bernoulli trial at stage j, which is either 1 or 0 depending on its success.

The reduction in computation time results from estimating the conditional probabilities $\mathrm{P}\{B_{i+1} \mid B_i\}$, which are not rare if the sets B_i are selected properly. Even more computational effort is saved by discarding paths that leave a set B_i, and which are therefore deemed unsuccessful. The optimal gain in computation time [15] is achieved if the sets are chosen such that[2]

$$\begin{aligned} M &= -\frac{1}{2} \ln(\mathrm{P}\{A\}) \\ \mathrm{P}\{B_{i+1} \mid B_i\} &= e^{-2} \\ R_i &\approx \frac{1}{\mathrm{P}\{B_{i+1} \mid B_i\}} = e^2 \end{aligned}$$

It should be noted that the optimal conditional probabilities as well as number of retrials do not depend on the model.

The numbers of retrials R_i can only be set approximately, because they have to be natural numbers. Apart from that, the optimal setting of the

[1] We assume $Thr_0 = -\infty$ and $Thr_{M+1} = \infty$ here to simplify notation.

[2] Later results of the same authors [14] recommend an alternative setting such that $\mathrm{P}\{B_{i+1} \mid B_i\} = \frac{1}{2}$, if it is possible to set the thresholds dense enough.

RESTART parameters for a given model and performance measure is not trivial. The first problem is that, following the equations above, the final result needs to be known before the simulation has been started. The second problem is how the importance function and especially the thresholds should be set. The problem of finding good thresholds is e.g. discussed in [11]. Finally, even if the optimal settings are known, the model structure might require to set e.g. different thresholds. However, even if the optimal efficiency might not be reached easily, experiences show that the technique works robustly for a wide range of applications [14,13].

Several variants of RESTART have been considered in the literature [1]. We follow the approach taken in [8,11], which can be characterized as *fixed splitting* and *global step* according to [1]. The first aspect corresponds to the number of trials into which a path is split when it reaches a higher level. The second issue governs the sequence in which the different trials are executed. Global step has the advantage to store fewer intermediate simulation states.

Following the presentation in [11], the steady-state value of our example measure $\mathrm{P}\{A\}$ is for a standard simulation given by

$$\mathrm{P}\{A\} = \lim_{T\to\infty} \frac{1}{T} \int_0^T \mathbf{1}_A(t)\,\mathrm{d}t$$

if we denote by $\mathbf{1}_A(t)$ the indicator variable that is either one or zero, depending on whether the current state of the simulation at time t is in A.

An estimator for this steady-state measure for a RESTART implementation needs correction factors that take into account the splitting. We adopt the method of [11], where *weights* ω are maintained during the simulation run, which capture the relative importance of the current path elegantly.

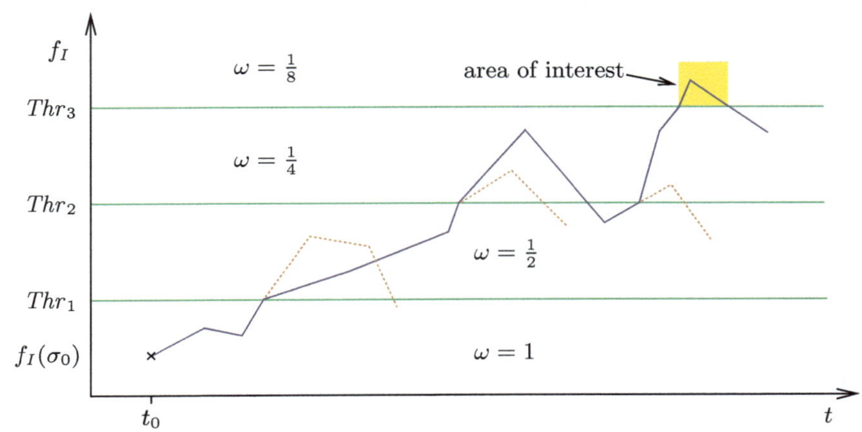

Fig. 2. Simulation runs in a RESTART algorithm

The weights are computed as follows: A simulation run starting from the initial state $\sigma_0 \in (B_0 \setminus B_1)$ has an initial weight of 1, because it is similar to a "normal" simulation run without splitting. Whenever the simulation path currently in level i crosses the border to an upper level u, the path is split into R_u paths, which are simulated subsequently. The weight is obviously divided by R_u upon splitting. Paths leading to a level $< i$ are discarded, except for the last one, which is followed further using the stated rules. The weight of the last path is multiplied by R_i when it leaves level i downwards. The weights of the previously discarded paths are thus taken back into consideration, to maintain an overall path probability of one. This procedure is repeated until the required result quality is achieved. This technique has the advantage of allowing "jumps of levels" over more than one threshold compared to the original method.

Fig. 2 depicts a sample evolution of a RESTART simulation with three thresholds. Every path that crosses the border of a threshold upwards is split into two paths in this example; the ones that drop below a threshold are canceled, unless they represent the last trial.

Based on the weight factors, an estimator for the steady-state probability of A is

$$\widehat{\mathrm{P}\{A\}} = \frac{1}{T} \int_0^T \omega(t)\, \mathbf{1}_A(t)\ \mathrm{d}t \tag{2}$$

with a large T. T counts in this context only the time spent in final paths, i.e. in the last path of each split.

3 RESTART for Extended Reward Measures

Approaches in the rare-event simulation literature estimate the probability of a rare state set A in transient or steady-state. This is however a significant restriction in the context of embedded systems evaluation and their respective SDES models. We apply the RESTART technique in the version described in [11] to the simulation of SDES models here, and extend it such that a complex SDES reward measure can be estimated.

Instead of estimating $\mathrm{P}\{A\}$, the goal is to obtain an estimation of a reward variable *rvar*. This extension is useful for all performance measures that significantly depend on rewards gained in areas of the state space which are only visited rarely. For simplicity of notation, we restrict ourself to one measure which is assumed to be analyzed in steady state.

3.1 Reward Measures

Reward variables are functions that return some value of interest from the stochastic process of a SDES model. This process $SProc = \{(\sigma(t), SE(t)), t \in \mathbb{R}^{0+}\}$ describes the state (σ) and possibly happening events (SE) at time t.

Reward variables describe combinations of a positive bonus or negative penalty associated to elements of the stochastic process. Two types of elements of such a reward variable have been identified in the literature [10]. This was based on the basic observation that the stochastic process of a discrete event system remains in a state for some time interval and then changes to another state due to an activity execution, which takes place instantaneously. The natural way of defining a reward variable thus includes **rate rewards** which are accumulated over time in a state, and **impulse rewards** which are gained instantaneously at the moment of an event.

We first introduce an intermediate function $R_{inst}(t)$. This value can be interpreted as the instantaneous reward gained at a point in time t. It is a generalized function in that it contains a Dirac impulse Δ if there is at least one impulse reward collected in t.

$$R_{inst}(t) = \underbrace{rrate\big(\sigma(t)\big)}_{\text{rate rewards}} + \underbrace{\Delta \cdot \sum_{se \in SE(t)} rimp(se)}_{\text{impulse rewards}} \tag{3}$$

We define the reward variable value in steady-state

$$rvar(SProc) = \lim_{x \to \infty} \frac{1}{x} \int_0^x R_{inst}(t)\,\mathrm{d}t \tag{4}$$

This leads to a simulation estimator $\widehat{rvar}$ in the sense of Equation (2).

$$\widehat{rvar} = \frac{1}{T} \int_0^T \omega(t)\, R_{inst}(t)\ \mathrm{d}t \tag{5}$$

where T is the (sufficiently large) maximum simulation time spent in final paths, and $R_{inst}(t)$ denotes the instantaneous reward gained at time t which is derived by the simulation, and $\omega(t)$ is the weight as managed by the algorithm and described in the previous section.

An importance function f_I for SDES states needs to be specified together with threshold values Thr_i that correspond to the different levels. There is no method known yet that obtains an importance function automatically for general models such as SDES (see e.g. the discussion in [11] and the references therein). In many model classes it is however easy to find one for the modeler. If for instance in a queuing model a high number of customers in a particular queue is seldom reached, lower numbers of the same value are natural thresholds. In the case of a reward measure defined as the probability that a certain number of tokens is exceeded in a place of a Petri net, an automatic method is for instance known [7, 8]. In the following we assume that thresholds are either specified by the user or obtained with a set of pre-simulations, as it is e.g. done in the implementations of SPNP [11] and TimeNET [8].

3.2 An Algorithm

The RESTART method for SDES models is started with RESTARTSIMULATION shown in Algorithm 3.2. Its input is the SDES model to be simulated, and a single reward variable *rvar* as a part of it. The initial state of the simulation is set with the initial SDES state and empty activity list (or event list). Accumulated reward and simulation time start with zero. The actual simulation function RESTARTPATH is called with level zero[3] and weight one. The call returns when the stop conditions are reached; an estimation of the performance measure is finally computed. It should be noted that SimTime is not the sum of all simulated time spans as in a standard simulation. It equals the simulation time spent in all final paths, compare Equation (5).

Algorithm 3.2 implements RESTARTPATH, the actual simulation of a path in the RESTART approach. The two functions are divided mainly to allow recursive execution of the latter. The main simulation loop is in part similar to a standard steady-state simulation, and only the differences are explained here. Its parameters include RESTART-specific level and weight as well as the complete state of the simulation itself, namely state, activity list and simulation time. This is necessary to start simulation paths at splitting points. The reward that is accumulated in a state and upon state change is multiplied by the weight factor ω, to compensate for the splitting procedure.

Level control according to the RESTART rules and Equation (1) is done at the end of the loop. If the next state is on a higher level lvl' than the current simulation state, the path is split into $R_{lvl'}$ paths that are started with recursive calls to RESTARTPATH. Each new path starts at the current new state, with a weight ω divided by the number of paths. The final state of the simulation that is reached with the last of the paths is followed thereafter by copying the returned state information.

RESTARTSIMULATION (SDES)

Input: SDES model with performance measure definition *rvar*
Output: estimated values of performance measures $rvar_i \in RV$

(∗ initializations ∗)
(∗ Set all state variables to their initial value ∗)
ActivityList $:= \emptyset$; $t_0 :=$ Reward$_{rvar} := 0$
(∗ call path simulation ∗)
$(\cdot, \cdot, \text{SimTime}) := \text{RESTARTPATH}(0, 1.0, \sigma_0, \text{ActivityList}, t_0)$
(∗ compute performance measure ∗)
$\text{Result}_{rvar} := \frac{\text{Reward}_{rvar}}{\text{SimTime}}$

Fig. 3. Startup of RESTART simulation

[3] We assume that $Level(\sigma_0) = 0$ for simplicity.

```
RESTARTPath (lvl, ω, σ, ActivityList, t)
-----------------------------------------------------------------
Input: Level lvl, Weight ω, state σ, activity list, time t
Output: Final state of the simulation: new (σ, ActivityList, t)

(* main simulation loop *)
repeat
    (* get new activities *)
    UpdateActivityList(σ, ActivityList, t)
    (* select executed activity *)
    (a, mode, t') := SelectActivity(ActivityList)
    Event := (a, mode)
    (* update performance measure rvar = (rrate, rimp, ·, ·) *)
    Reward_rvar += ω * ((t' − t) * rrate(σ) + rimp(Event))
    (* execute state change *)
    t := t'; σ := Exec(Event, σ)
    (* RESTART level control *)
    lvl' := Level(σ)
    if lvl' > lvl then (* split *)
        for i = 1 ... R_lvl' do (σ', ActivityList', t') :=
                RESTARTPath(lvl', ω/R_lvl', σ, ActivityList, t)
        (* Continue the final path *)
        σ := σ'; ActivityList := ActivityList'; t := t'
until (lvl' < lvl) or (stop condition reached, e.g. t ≥ MaxSimTime)
return (σ, ActivityList, t)
```

Fig. 4. Main RESTART algorithm for steady-state simulation

The main simulation loop repeats until a new state belongs to a lower level; the current path is then aborted by a return to the upper level of recursion. In addition to that, some stop condition depending on the simulation time or the accuracy achieved so far is used. Corresponding formulas for the estimation of result variance in a RESTART simulation can be found in the literature, see e.g. [14,13].

4 Conclusion

In this paper we have shown how the RESTART method for rare-event simulation can be applied to stochastic discrete-event systems containing more general performance measures than only the probability of one rare event. We extend the technique to general reward measures containing impulse and rate rewards.

References

1. M. J. Garvels and D. P. Kroese, "A comparison of RESTART implementations," in *Proc. 1998 Winter Simulation Conference*, 1998.
2. P. Glasserman, P. Heidelberger, P. Shahabuddin, and T. Zajic, "Multilevel splitting for estimating rare event probabilities," *Operations Research*, vol. 47, pp. 585–600, 1999.
3. P. W. Glynn and D. L. Iglehart, "Importance sampling for stochastic simulations," *Management Science*, vol. 35, no. 11, pp. 1367–1392, Nov. 1989.
4. C. Görg, E. Lamers, O. Fuš, and P. Heegaard, "Rare event simulation," Computer Systems and Telematics, Norwegian Institute of Technology, Tech. Rep. COST 257, 2001.
5. A. Goyal, W. C. Carter, E. de Souza e Silva, S. S. Lavenberg, and K. S. Trivedi, "The system availability estimator," in *Proc. 16th Symp. Fault-Tolerant Computing*, Vienna, Austria, 1986, pp. 84–89.
6. P. Heidelberger, "Fast simulation of rare events in queueing and reliability models," *ACM Transactions on Modeling and Computer Simulation (TOMACS)*, vol. 5, no. 1, pp. 43–85, 1995.
7. C. Kelling, "Rare event simulation with RESTART in a Petri net modeling environment," in *Proc. of the European Simulation Symposium*, Erlangen, 1995, pp. 370–374.
8. C. Kelling, "A framework for rare event simulation of stochastic Petri nets using RESTART," in *Proc. of the Winter Simulation Conference*, 1996, pp. 317–324.
9. W. Obal and W. Sanders, "Importance sampling simulation in UltraSAN," *Simulation*, vol. 62, no. 2, pp. 98–111, 1994.
10. W. H. Sanders and J. F. Meyer, "A unified approach for specifying measures of performance, dependability, and performability," in *Dependable Computing for Critical Applications*, Dependable Computing and Fault-Tolerant Systems, A. Avizienis and J. Laprie, Eds. Springer Verlag, 1991, vol. 4, pp. 215–237.
11. B. Tuffin and K. S. Trivedi, "Implementation of importance splitting techniques in stochastic Petri net package," in *Computer Performance Evaluation, Modelling Techniques and Tools — 11th Int. Conf., TOOLS 2000*, Lecture Notes in Computer Science, C. U. S. Boudewijn R. Haverkort, Henrik C. Bohnenkamp, Ed., vol. 1786. Schaumburg, IL, USA: Springer Verlag, 2000, pp. 216–pp.
12. M. Villén-Altamirano and J. Villén-Altamirano, "RESTART: A straightforward method for fast simulation of rare events," in *Proc. Winter Simulation Conference*, 1994, pp. 282–289.
13. M. Villén-Altamirano and J. Villén-Altamirano, "Analysis of RESTART simulation: Theoretical basis and sensitivity study," *European Transactions on Telecommunications*, vol. 13, no. 4, pp. 373–385, 2002.
14. M. Villén-Altamirano and J. Villén-Altamirano, "Optimality and robustness of RESTART simulation," in *Proc. 4th Workshop on Rare Event Simulation and Related Combinatorial Optimisation Problems*, Madrid, Spain, Apr. 2002.
15. M. Villén-Altamirano, J. Villén-Altamirano, J. Gamo, and F. Fernández-Cuesta, "Enhancement of the accelerated simulation method RESTART by considering multiple thresholds." in *Proc. 14th Int. Teletraffic Congress*. Elsevier Science Publishers B. V., 1994, pp. 797–810.

High Performance Low Cost Multicore NoC Architectures for Embedded Systems

Dietmar Tutsch and Günter Hommel

Real-Time Systems and Robotics
Technische Universität Berlin, Germany
{dietmart,hommel}@cs.tu-berlin.de

1 Introduction

Many embedded systems require high performance computing. The advent of multicore processors on a single chip is a big step forward for such compute-intensive applications. For mass products, not only the performance but also the cost of those systems is essential.

To allow cooperating cores on such a multicore processor, an appropriate communication structure must be provided. In case of a low number of cores (e.g. a dual core processor), a shared bus may be sufficient. But in future, hundreds or even thousands of cores will collaborate on a single chip. Then, more advanced network topologies will be needed. Many topologies are proposed for these so called networks-on-chips (NoCs) [1, 2, 3]. Multistage interconnection networks (MINs) belong to them [4, 5].

To map the communication demands of the cores onto predefined topologies like MINs, but also other topologies like meshes, busses, etc., Bertozzi et al. [3] developed a tool called NetChip (consisting of SUNMAP and xpipes. This tool provides complete synthesis flows for NoC architectures.

Guerrier and Greiner [4] established MINs with fat tree structure using Field Programmable Gate Arrays (FPGAs). They called this on-chip network with particular router design and communication protocol Scalable, Programmable, Integrated Network (SPIN). Its performance for different network buffer sizes was compared.

Alderighi et al. [5] used MINs with the Clos structure. Multiple parallel Clos networks connect the inputs and outputs to achieve fault tolerance abilities. Again, FPGAs serve as basis for realization.

Previous papers only considered unicast traffic in the NoC. But it is obvious that multicore processors also deal with multicast traffic. For instance, if a core changes a shared variable that is also stored in the cache of other cores, multicasting the new value to the other cores keeps them up to date. Thus, multicast traffic builds a non-negligible part of the traffic.

Günter Hommel and Sheng Huanye (eds.), Embedded Systems Modeling – Technology, and Applications, 53–62.

In this paper, a new network-on-chip architecture called multilayer multistage interconnection network (MLMIN) is introduced that particularly applies to multicasting in packet switched multistage interconnection networks for multicore processors. Besides investigating network traffic of constant size packets [6], particularly traffic consisting of packets of varying size is considered. Varying size packets lead to a more complex performance evaluation and buffer cost comparison.

The paper is organized as follows. The NoC architecture of multilayer multistage interconnection networks is described in Section 2. To evaluate the new network architecture, its cost is determined (Section 3) as well as its performance (Section 4). These results are compared to other NoC architectures. Section 5 summarizes and gives conclusions.

2 Network-on-Chip Architectures

The NoC architecture of multilayer multistage interconnection networks presented in this paper is based on MINs with the banyan property. Multistage Interconnection Networks are dynamic networks which are based on switching elements (SE). SEs are arranged in stages and connected by interstage links. The link structure and amount of SEs characterizes the MIN.

2.1 Banyan-like MIN

MINs with the banyan property are networks where a unique path exists from an input to an output. Such MINs of size $N \times N$ consist of $c \times c$ switching elements with $N = c^n$ and $n, c, N \in \mathbb{N}$. The number of stages is given by n. To achieve synchronously operating switches, the network is internally clocked. A message to be transferred is divided into packets. The packets consist of a header representing the routing tag and of a payload field filled with the corresponding part of the message.

Because packet switching is applied buffers can be introduced. In each stage k ($k \in \mathbb{N}$ and $0 \leq k \leq n-1$), there is a FIFO buffer of $M(k)$ bytes in front of each switch input. The packets may be of varying size f_i but smaller or equal to the smallest buffer size: $f_i \leq \min_k\{M(k)\}$. As already mentioned, when applying MINs for multicore processors, this will most likely raise multicast traffic. Multicasting can be performed by copying the packets within the $c \times c$ SEs instead of copying them before they enter the network. This scheme is called message replication while routing (MRWR). Comparing the packet density in the stages in case of replication while routing shows that the greater the stage number, the higher is the amount of packets. In other words: there are usually much more packets in the last stages due to replication than in the first ones.

2.2 Multilayer MIN

Multilayer multistage interconnection networks consider the multicast traffic characteristics in multicore processors. As mentioned in Section 2.1, the amount of packets increases from stage to stage due to packet replication. Thus, more switching power is needed in the last stages compared to the first stages of a network.

To supply the network with the required switching power, the new NoC structure presented in this paper replicates the number of layers in each stage. The factor with which the number of layers is increased is called growth factor G_F ($G_F \in \mathbb{N}\backslash\{0\}$). Figure 1(a) shows an 8×8 MLMIN (3 stages) with growth factor $G_F = 2$ in lateral view. Here, the number of layers is doubled each stage

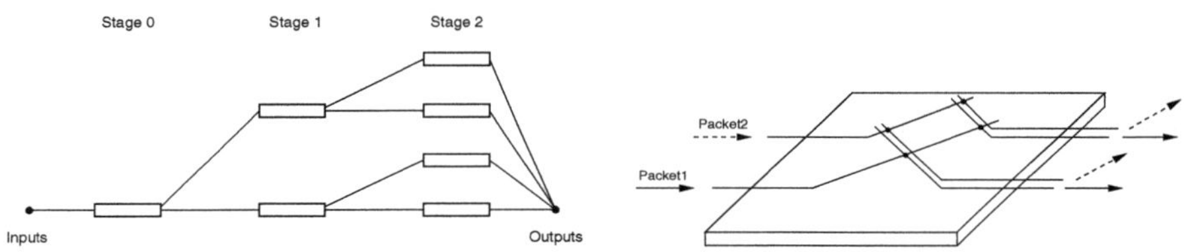

(a) Topology (lateral view) (b) SE with unblocked broadcast

Fig. 1. Multilayer multistage interconnection network ($G_F = 2$)

and each switching element has twice as much outputs as inputs. Consider for instance that 2×2 SEs are used. Such an architecture ensures that even in case of two broadcast packets at the inputs all packets can be sent to the outputs (if there is buffer space available at the succeeding stage). For instance, one packet is broadcasted to the lower layer and the other one to the upper layer. Figure 1(b) shows such a scenario. As can be seen in the figure, the SE architecture of MLMINs slightly differs from regular SEs: The outputs are replicated by the growth factor G_F to connect the switch to all G_F layers.

Replicating the number of layers in each stage avoids unnecessary layer replications in the first stages as in case of the replicated MINs. Choosing $G_F = c$ ensures that no internal blocking occurs in an SE, even if all SE inputs broadcast their packets to all SE outputs. Nevertheless, blocking may still occur at the network output depending on the number R of accepted packets per clock cycle.

A drawback of the new architecture arises from the exponentially growing number of layers for each further stage. The more network inputs are established, the more stages and the more layers result. To limit the number of layers and therefore the amount of needed chip area, two options are considered: starting the replication in a more rear stage and/or stopping further layer replication if a given number of layers is reached.

For the first option, a stage number in which replication starts is defined by G_S ($0 \leq G_S < n$ and $G_S \in \mathbb{N}$). Stopping further layer replication if a given number G_L of layers is reached also reduces the network complexity ($G_L \in \mathbb{N}\backslash\{0\}$). It prevents exponential growth beyond reasonable limits in case of large networks. Both presented options can be combined to further reduce network complexity. Such a network is determined by parameters G_S (start of replication), G_F (growth factor), and G_L (layer limit).

3 Network Hardware Cost

An important issue in designing networks-on-chips is given by their hardware cost determined for instance by the required chip area. In this paper, to be independent from the current chip technology, the cost of the presented networks are compared by determining the number of crosspoints within each network and the number and size of their buffers. Setting both items in an appropriate relation and adding them gives the overall network cost.

3.1 Crosspoint Cost

To determine the crosspoint cost, their number within the network is calculated. It is given by the number of crosspoints within a switching element and by the number of switching elements within the network.

SEs of banyan-like MINs consist of c^2 crosspoints if $c \times c$ SEs are used. That means the crosspoint cost P_{MIN} of an $N \times N$ MIN, which consists of $n = \log_c N$ stages with N/c SEs in each stage, results in

$$P_{MIN} = n \cdot \frac{N}{c} \cdot c^2 = n \cdot c^{n+1} \tag{1}$$

Replicated MINs are established by L layers of MINs, a demultiplexer at each input, and a multiplexer at each output. Each demultiplexer is composed of L crosspoints to distribute the incoming packets among the L layers. A multiplexer is also composed of L crosspoints to collect the packets from the L layers. Thus, the crosspoint cost P_{RepMIN} of a replicated MIN are given by

$$P_{RepMIN} = L \cdot n \cdot c^{n+1} + L \cdot c^n + L \cdot c^n = L \cdot c^n \cdot (n \cdot c + 2) \tag{2}$$

The crosspoint cost of multilayer multistage interconnection networks are also mainly determined by the number and crosspoint cost of the switching elements. As shown in Figure 1(b), the structure of SEs that allow increasing number of stages in the succeeding stage is slightly changed to regular SEs of banyan-like MINs. Each output is replicated by G_F. That means $c \times c$ SEs of such a structure consist of c inputs and $G_F \cdot c$ outputs resulting in

$$P_{MLSE} = c^2 \cdot G_F \tag{3}$$

crosspoints. An MLMIN in which layer growing starts at Stage G_S ($G_S > 0$) and the number of layers is not limited consists of $G_S - 1$ stages of regular SEs, followed by $n - G_S$ stages in which SEs according to Figure 1(b) are used (the number of layers is replicated in each stage by G_F), and a last stage of $G_F^{n-G_S}$ regular SEs (no output replication is needed in last stage; see Figure 1(a)). Additionally, a multiplexer connects the $G_F^{n-G_S}$ layers at each network output. The MLMIN crosspoint cost add to

$$
\begin{aligned}
P_{MLMINnolimit} &= \frac{N}{c} \cdot \left((G_S - 1) \cdot c^2 + c^2 \cdot G_F \cdot \sum_{i=0}^{n-1-G_S} G_F^i + G_F^{n-G_S} \cdot c^2 \right) \\
&\quad + N \cdot G_F^{n-G_S} \\
&= c^n \cdot \left(c \cdot \left(G_S - 1 + G_F \frac{G_F^{n-G_S} - 1}{G_F - 1} + G_F^{n-G_S} \right) + G_F^{n-G_S} \right)
\end{aligned} \tag{4}
$$

where the sum represents a geometric series and is replaced by its solution.

The crosspoint cost of an MLMIN with a limited number G_L of layers can be obtained by adapting Equation (4). First, the multiplexer cost are removed (they are added again later on). Then, the equation gives the crosspoint cost of the first network part till the growing of the layers stops at stage x (n within the brackets of Equation (4) must be replaced by x). x represents the number of the first stage that has equal number of layers as the previous one. The remaining $n - x$ network stages consist of G_L layers of regular SEs, followed by multiplexers connecting the layers.

$$
\begin{aligned}
P_{MLMINlimit} &= N \cdot \left(c \cdot \left(G_S - 1 + G_F \frac{G_F^{x-G_S} - 1}{G_F - 1} + G_F^{x-G_S} \right) \right) \\
&\quad + \frac{N}{c} \cdot (n - x) \cdot G_L \cdot c^2 + N \cdot G_L
\end{aligned} \tag{5}
$$

Growing by factor G_F stops at the stage at which the number of layers reaches the limit G_L:

$$G_F^{(x-G_S)} = G_L \tag{6}$$

Thus, stage x is determined by

$$x = G_S + \log_{G_F} G_L \tag{7}$$

with $G_L \in \left\{ G_F^i \;\middle|\; i \in \mathbb{N}\backslash\{0\} \wedge i \leq n - G_S \right\}$. Equation (5) results in

$$
\begin{aligned}
P_{MLMINlimit} &= c^n \cdot \left(c \cdot \left(G_S - 1 + G_F \frac{G_L - 1}{G_F - 1} \right.\right. \\
&\quad \left.\left. + (n + 1 - G_S - \log_{G_F} G_L) \cdot G_L \right) + G_L \right)
\end{aligned} \tag{8}
$$

Equation (8) is also valid for MLMINs with no limited number of layers if G_L is set to $G_L = G_F^{n-G_S}$.

3.2 Buffer Cost

The buffer cost of a network are determined by the number of buffers and their sizes $M(k)$. In the following, it is assumed that all buffers are of equal size M.

Specifying buffer cost in simply adding the buffer sizes all over the network would not be sufficient. Buffer cost must be related to crosspoint cost to allow to sum them up. For instance, considering hardware cost in terms of approximately needed transistors solves the problem.

An assumption could be that a crosspoint is realized by a single transistor which is responsible to close or open the crosspoint connection. On the other hand, to store one bit, usually two transistors are necessary (dynamic RAM): one of them to establish a connection to the data line of the memory and one realizes the capacitor. That means to store one byte consisting of 8 bit, $8 \cdot 2 = 16$ transistors are needed. To transfer one byte assuming that 8 bit are transferred in parallel, 8 parallel crosspoints (transistors) are needed. The relation V between one byte memory and parallel crosspoints results in $V = \frac{16}{8} = 2$ for this example. In general, the relation is given by $V = \frac{16}{b}$ where b represents the number of bits that are transferred in parallel in the interconnection network. Today's architectures usually use $b = 32$ or $b = 64$.

Replicated MINs consist of L layers of MINs. Each $N \times N$ MIN is established by n stages of SEs. All N SE inputs of a stage are connected to a buffer of M bytes. Therefore, the buffer cost of a replicated MIN is given by

$$B_{RepMIN} = V \cdot M \cdot L \cdot N \cdot n \tag{9}$$

This equation is also valid for banyan-like MINs if the number of layers is set to $L = 1$.

The number of buffers of a MLMIN with no layer limit is determined by $G_S - 1$ stages where no growing occurs (resulting in just one layer) and by the remaining $n+1-G_S$ stages where layers grow with factor G_F. Each layer of a stage consist of N buffers (due to N SE inputs at each stage and layer) of size M leading to buffer cost of a MIN with no layer limit of

$$\begin{aligned} B_{MLMINnolimit} &= V \cdot M \cdot N \cdot \left(G_S - 1 + \sum_{i=0}^{n-G_S} G_F^i \right) \\ &= V \cdot M \cdot N \cdot \left(G_S - 1 + \frac{G_F^{n+1-G_S} - 1}{G_F - 1} \right) \end{aligned} \tag{10}$$

MLMINs with a layer limit of G_L consist similar to unlimited MLMINs of $G_S - 1$ stages where no growing occurs. $x + 1 - G_S$ stages deal with layer growing with x defined as in Equation (7). $n-x$ stages remain with maximum number of layers G_L. Thus, buffer cost of MLMINs with a layer limit result in Equation (11). This equation is also valid for MLMINs with no limited number of layers if G_L is set to $G_L = G_F^{n-G_S}$.

$$\begin{aligned} B_{MLMINlimit} &= V \cdot M \cdot N \cdot \left(G_S - 1 + \sum_{i=0}^{x-G_S} G_F^i + (n-x) \cdot G_L \right) \\ &= V \cdot M \cdot N \cdot \left(G_S - 1 + \frac{G_L \cdot G_F - 1}{G_F - 1} \right. \\ &\qquad \left. + (n - G_S - \log_{G_F} G_L) \cdot G_L \right) \end{aligned} \tag{11}$$

4 Network Performance

In the first part of this section, the performance of various NoC architectures based on MINs is compared. Just packets of constant size f are offered to the networks to achieve better comparability. The second part of this section concentrates on the performance of MLMINs in which packets of varying size are transferred. The normalized throughput S_o at the outputs and the mean delay time d_{tot} serve as performance measures. A multicasting traffic is assumed where all possible combinations of destination core addresses for each packet entering the network are equally distributed. Furthermore, equal traffic rate is offered to all network inputs. Performance results are obtained by simulation. Statistical analysis of the results is performed by *Akaroa* [7]. As simulator termination criteria, the estimated precision was set to 2% and the confidence level to 95%.

4.1 Packets of Constant Size

In a first step, the presented network architectures are investigated in case of network traffic that consists of packets of constant size f. All results are obtained using buffers that are able to store a single packet: $M = f$. In consequence, network cost are given dependent on the buffer-to-crosspoint relation V and the packet size f. To give an example: if $b = 32$ bits are transferred in parallel, $V = 0.5$ results (see Section 3.2). If further on, packets are assumed to be of size $f = 40$ bytes (networks-on-chips usually use small packet sizes of a few bytes), a factor of $V \cdot f = 20$ results to relate buffer hardware cost for a single packet to crosspoint cost. Figure 2 shows the performance of 16×16 networks consisting of 2×2 SEs. Multiple acceptance of $R = 4$ packets per network output and per clock cycle is assumed. The layer to which a packet is send is chosen randomly. Also, conflicts among packets for resources (SE outputs, buffers, layers) are solved by random. The networks operate in store and forward switching and global backpressure.

Because 2×2 SEs are used, packets can be doubled at most at each stage even in case of broadcasting the packets to all network outputs. Therefore, doubling the number of layers in each stage is a sufficient growing. That leads to a growth factor of $G_F = 2$ with growing to be started at stage $G_S = 1$ and no layer limit. The resulting 16×16 MLMIN consists of one layer at Stage 0, two layers at Stage 1, four layers at Stage 2, and eight layers at Stage 3. Such a network is named *1248* in the following figures (The digits of the legend refer to the number of layers at Stage 0, Stage 1, etc).

Figure 2 compares Network *1248* with less expensive ones that consist just of four layers in the last stage. Additionally, it is compared with replicated MINs. Network *1248* performs in terms of throughput best in relation to the

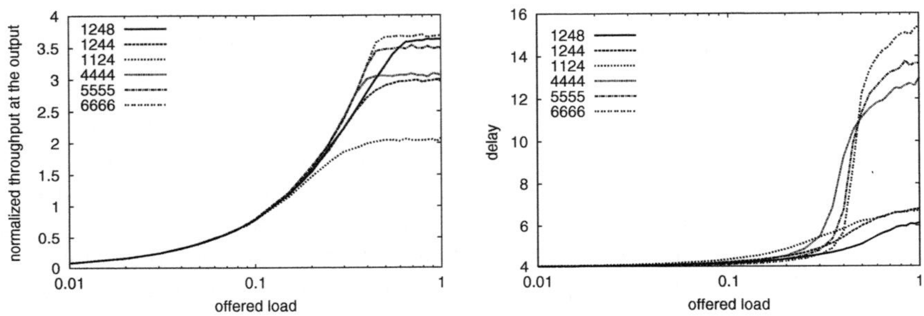

Fig. 2. Performance comparison of different architectures

networks with four layers in last stage. Networks (replicated MINs with five and six layers) that catch up with its throughput suffer from much higher delays and from higher costs if $V \cdot f > 1$ is assumed (e.g. $V \cdot f = 20$). Costs of the networks are given in Table 1(a). Again, Network *1248* is the best solution.

Table 1. Cost of architectures

(a) Packets of constant size

Network	Parameter			Cost	
	G_S	G_F	G_L	Crossp.	Buffer
1248	1	2	–	832	$240 \cdot V \cdot f$
1244	1	2	4	512	$176 \cdot V \cdot f$
1124	2	2	–	416	$128 \cdot V \cdot f$
4444	rep. MIN: $L = 4$			640	$256 \cdot V \cdot f$
5555	rep. MIN: $L = 5$			800	$320 \cdot V \cdot f$
6666	rep. MIN: $L = 6$			960	$384 \cdot V \cdot f$

(b) Packets of varying size

Network	Cost	
	Crossp.	Buffer
MLMIN1248	832	$48000 \cdot V$
MIN1248	128	$48000 \cdot V$
MIN1111	128	$12800 \cdot V$

4.2 Packets of Varying Size

In the following, multilayer multistage interconnection networks are examined in case of network traffic that consists of multicast packets of varying size. For instance for multiprocessor systems, a packet size of about 100 bytes is reasonable. Such a size is also reasonable for NoCs in multicore processors and thus will deal as an example as follows: It is assumed that packets of a size uniformly distributed between 5 bytes and 150 bytes are offered to the NoC. The traffic in 16×16 networks consisting of 2×2 SEs is investigated.

Figure 3 compares an MLMIN with two kinds of banyan-like MINs. The MLMIN (named *MLMIN1248*) deals with a growth factor of $G_F = 2$ with growing to be started at stage $G_S = 1$ and no layer limit. All buffers are of size $M = 200$ bytes. Network *MIN1248* represents a banyan-like MIN. To

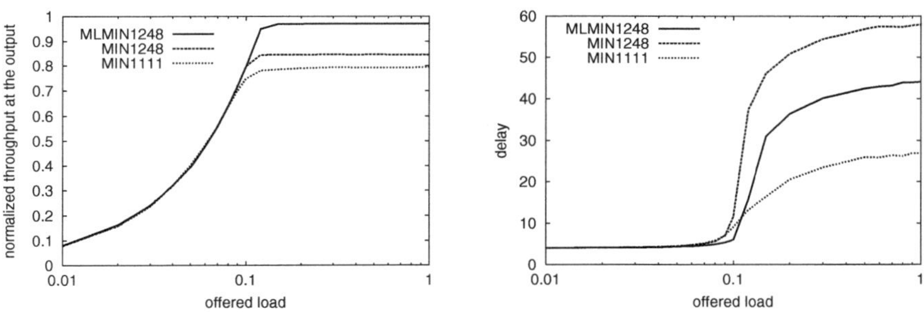

Fig. 3. Traffic of varying size packets

keep the overall buffer size equal to *MLMIN1248*, all corresponding MLMIN buffers of the different layers are merged to a single buffer at the single layered banyan-like MIN. Consequently, a MIN with buffer size $M(0) = 200$ bytes at Stage 0, $M(1) = 400$ bytes at Stage 1, $M(2) = 800$ bytes at Stage 2, and $M(3) = 1600$ bytes at Stage 3 results. The third network (named *MIN1111*) is equal to *MIN1248* except that all buffers are of size $M = 200$ bytes. Single acceptance ($R = 1$) at network outputs is performed.

As depicted in Figure 3, the MLMIN shows the highest throughput due to more switching power and less blockings. Even single acceptance cannot countervail the better performance. The MLMIN also outperforms Network *MIN1248* in delay times. Therefore, merging all buffers to a single layer gives no alternative to MLMINs, particularly because no significant cost difference exists (Table 1(b)). Network *MIN1111* attracts by its low delay times due to less and smaller buffers. Less and smaller queues result leading to lower waiting times. On the other hand, *MIN1111* suffers from its low throughput.

5 Conclusions

With multilayer multistage interconnection networks, a very efficient network-on-chip topology for multicast traffic in multicore processors is introduced. This paper describes how MLMINs are established and what parameters determine such an architecture. Performance and cost of MLMINs are calculated and compared to other proposed NoC topologies.

It turned out that the MLMIN architecture shows a better performance or at least similar performance in terms of throughput if network costs are equal. The delay times of MLMINs heavily undercut those of replicated MINs. Thus, multicore architectures using MLMINs are obvious candidates for the implementation of compute-intensive embedded systems.

References

1. William J. Dally and Brian Towles. Route packets, not wires: On-chip interconnection networks. In *Proceedings of Design Automation Conference (DAC 2001)*, pages 684–689, 2001.
2. L. Benini and G. De Micheli. Networks on chips: A new SoC paradigm. *IEEE Computer*, 35(1):70–80, 2002.
3. Davide Bertozzi, Antoine Jalabert, Srinivasan Murali, Rutuparna Tamhankar, Stergios Stergiou, Luca Benini, and Giovanni De Micheli. NoC synthesis flow for customized domain specific multiprocessor systems-on-chip. *IEEE Transactions on Parallel and Distributed Systems*, 16(2):113–129, February 2005.
4. Pierre Guerrier and Alain Grenier. A generic architecture for on-chip packet-switched interconnections. In *Proceedings of IEEE Design Automation and Test in Europe (DATE 2000)*, pages 250–256. IEEE Press, 2000.
5. Monica Alderighi, Fabio Casini, Sergio D'Angelo, Davide Salvi, and Giacomo R. Sechi. A fault-tolerant FPGA-based multi-stage interconnection network for space applications. In *Proceedings of the First IEEE International Workshop o Electronic Design, Test and Applications (DELTA'02)*, pages 302–306, 2002.
6. Dietmar Tutsch and Günter Hommel. Multilayer multistage interconnection networks. In *Proceedings of 2003 Design, Analysis, and Simulation of Distributed Systems (DASD 2003); Orlando*, pages 155–162. SCS, April 2003.
7. G. Ewing, K. Pawlikowski, and D. McNickle. Akaroa2: Exploiting network computing by distributing stochastic simulation. In *Proceedings of the European Simulation Multiconference (ESM'99); Warsaw*, pages 175–181. SCS, 1999.

An Analyzable On-Chip Network Architecture for Embedded Systems

Daniel Lüdtke, Dietmar Tutsch, and Günter Hommel*

Real-Time Systems and Robotics
Technische Universität Berlin, Germany
{dluedtke, dietmart, hommel}@cs.tu-berlin.de

1 Introduction

The increasing integration level of modern and forthcoming Integrated Circuits (ICs) allows the implementation of complex systems on a single chip (System-on-Chip – SoC) [1]. To reduce time to market in the design flow of chip development prefabricated components (known as Intellectual Property – IP) are used. IP cores can be CPUs, memory blocks, signal processing units, etc. Simplified, the design of new complex customized ICs often represents a composition of IP cores. The main task of a chip designer is the selection, parameterization, and interconnection of the different cores.

Currently, the interconnection between the IP cores are realized with either dedicated point-to-point interconnections or standardized system buses. Dedicated point-to-point connections are only manageable and economically feasible in smaller systems. With increasing complexity, it is not possible to connect every core with dedicated wires. Buses, on the other hand, are an example for shared communication resources. The system design with a standardized bus and IP cores with an interface to the bus in question, becomes much simpler. Examples for on-chip buses are IBM Core Connect, AMBA-Bus by ARM, and the VCI-Standard by the VSIA. However, as a drawback, buses are not scalable for larger designs. The communication between cores becomes the performance bottleneck of the system.

SoC design tends to replace buses with packet-switched interconnection networks [5]. The architecture of switched on-chip networks are similar to

* The Deutsche Forschungsgemeinschaft (DFG) funded parts of this work under grant Ho 1257/22-1.

Günter Hommel and Sheng Huanye (eds.), Embedded Systems Modeling – Technology, and Applications, 63–72.

interconnection networks used in high performance parallel computers or in switching fabrics for Ethernet routers. Hence, it seems reasonable to adapt the well-known methods for the investigation of interconnection networks to this new research area. In particular, the methods of performance evaluation of on-chip networks are the focus of this paper. Actually, performance measures are usually obtained by simulation (e.g. [17]). Only some inaccurate analytical models have been proposed for the performance evaluation of on-chip network architectures (for instance: [6]).

Due to the lack of powerful and accurate performance evaluation methods, most proposals for on-chip networks are based on regular topologies (meshes in the majority of cases). Nevertheless, many applications that are realized as systems on a chip have no regular or symmetric communication requirements (an MPEG4 decoder for example [7]). In these cases, a customized network topology would lead to better performance with much less overhead in terms of area cost on chip and power consumption of the system [11].

The remainder of this paper presents an on-chip network architecture that is designed for high-speed SoC-designs. A further design goal is an architecture that could simply be transferred into an analytical model for stochastic performance evaluation. Parallel to the implementable network description, a cycle-accurate stochastical network model, based on discrete-time Markov chains, will be derived. In cooperation with analytical models for the estimation of power consumption and floor space requirements [3] this new performance model of on-chip networks will improve the design process of Systems-on-Chip, especially for embedded systems.

In the next section, the proposed network architecture is presented in detail. Sect. 3 gives a short sketch of deriving a stochastical Markov-based model from the register level description of the network architecture. Sect. 4 outlines possible application areas of the stochastical model and Sect. 5 summarizes and gives conclusions.

2 On-Chip Network Architecture

Many different proposals for on-chip network architectures have been published. They differ in details such as topology, switching scheme, flow control, arbitration, buffer arrangement, etc. The network architecture introduced below is designed in terms of enhanced flexibility, scalability, simplicity, and applicability in high-speed designs. The network architecture supports static networks like meshes and dynamic networks like multistage interconnection networks (MINs). All elements of the network are parmeterizable; for example buffer sizes and the number of inputs and outputs. Hence, it is possible to build a great variety of topologies like the irregular dynamic network depicted in Fig. 1. Switching, flow-control, and arbitration schemes are chosen to be powerful but simple enough for the transition to an accurate stochastic model.

As a starting point, the ×pipes Lite architecture [13] was evaluated and modified to meet the mentioned design goals. ×pipes Lite is a SystemC library of modules for building network interfaces, routers, and pipelined links to build arbitrary networks for on-chip applications.

The following sections describe the adapted network architecture that represents a good compromise between the needs of analysis and implementation in detail.

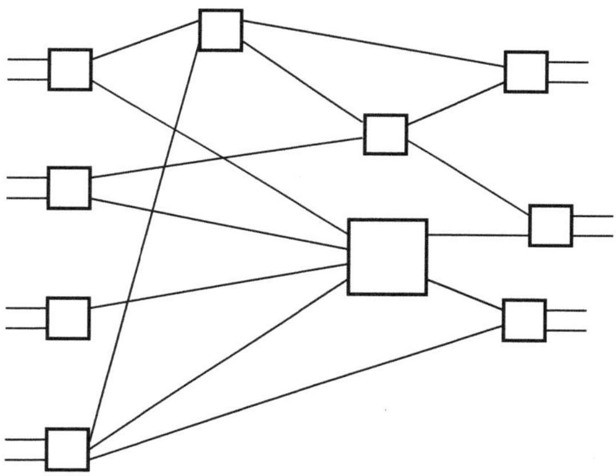

Fig. 1. Irregular dynamic network built with routers of different sizes

2.1 Components

Figure 2 depicts the register level structure of a very small network where all components of the proposed architecture are present. The main elements are routers, traffic sources, traffic destinations, and links.

A *router* is built with buffers, a switch, and the arbitration logic. All buffers in a router have to store one or several entire packets. The architecture supports input, output, or combined input/output buffering (as shown in Fig. 2). A crossbar or multiplexers implement the switch in the center of the router. It dispatches the packet to the destined router output according to the routing scheme. Conflicts, caused by several packets concurrently destined to the same output, are solved by the arbitration logic. It assigns packet priority in accordance with the flow control scheme. In an uncongested network, this router has a latency of two cycles.

IP cores are connected to the network by network interfaces. They are responsible for packetization (including the generation of a packet header with destination addressing) and synchronization. For further discussions, the interface is split up in a transmitting part: the traffic *source*; and a receiving part: the *destination*. Both are built with a buffer to store exactly one packet.

Links between components (sources, destinations, routers) are unidirectional and subdivided by pipeline registers. This ensures a high operating frequency of the network. The number of pipeline registers depends on the final layout of the resulting IC. Every pipeline register has the capacity of a single flit (flow control unit). Packets are divided into flits in such a way, that one flit can proceed from a buffer to another in a single cycle. It is assumed that all packets are of equal size.

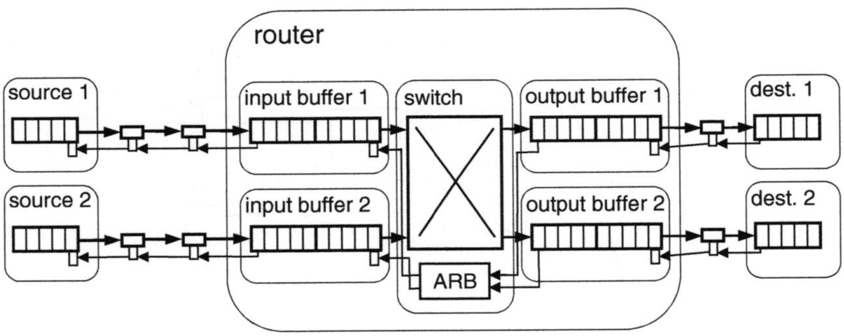

Fig. 2. Register level structure of a small network

2.2 Routing

A routing specific scheme is not predefined in the architecture description. Due to the freedom of choice in topology for the application area in question, a preselected routing scheme would be hindering. In principle, all distributed routing schemes (the routing decision is made within the routers) can be used in the final design.

The network supports both static and adaptive routing algorithms if redundant paths are available.

2.3 Switching Scheme

As mentioned before, packets are divided into smaller units (flits) which advance through the network in a pipelined manner. A buffer in the routers only accepts a packet if this packet can be stored in the buffer completely. Otherwise, the packet will be rejected. The packet is stored in a buffer until the next buffer in the path of the packet is ready to accept it. This can possibly be delayed due to output conflicts or congestion in routers ahead. If the output becomes available again, the packet will immediately be trans-transmitted even if the packet is not received completely. This switching scheme is called partial cut-through in contrast to ×pipes Lite where wormhole switching is realized.

Wormhole switching has the advantage of needing fewer buffers because it is not required that a buffer must store a whole packet. Therefore, it is the most commonly used flow-control scheme in implemented applications. The drawback of wormhole switching is the increased occurrence of congestion in the network and the possibility of deadlocks. Deadlock-avoidance is often realized by restrictions to the routing freedom [18] but that disagrees with the demand of a free choice of a routing scheme (Sect. 2.2).

2.4 Flow Control Scheme

The on-chip network is designed to prevent packet loss within the network. This means that a packet that has entered the network will eventually reach its destination. As a result, no acknowledgement messages have to be exchanged between the communicating cores.

Packet loss avoidance between different network components is realized with a simple ACK-protocol with Go-Back-N semantics. Every buffer transmits its first packet in its queue continuously. When the packet-header-flit reaches the next buffer and the packet is accepted, an ACK-bit is transmitted back to the source buffer (see again Fig. 2). After a round-trip delay of $2 \cdot \#(\textit{pipeline registers}) + 2$ the ACK can be evaluated by the source buffer. If the ACK is missing, the source buffer retransmits the packet. This protocol assumes that packets are divided into more flits than the maximum round-trip delay.

Arbitration in the routers is solved randomly for modeling purposes. In the final implementation, randomness will be replaced by round robin or priority-based arbitration schemes, depending on the application.

2.5 Multicast

Other on-chip network architectures only consider unicast traffic. However, in many applications multicast traffic occurs quite often. For instance, if one CPU-core changes a shared variable that is also stored in the cache of other CPU-cores, multicasting the new value to the other cores keeps them up to date. It is preferable, to support multicast in hardware otherwise multicast messages have to be injected into the network several times.

The hardware support for multicast is realized simply by copying the packets within the routers. This scheme is called message replication while routing (MRWR). Packets are copied as late as possible to optimally reduce the number of packets in the network, which leads to less congestion.

3 Stochastic Model Generation

To investigate how an on-chip network behaves based on the chosen network topology, buffer sizes, packet size, etc., a corresponding stochastic model will be designed for performance evaluation. Instead of modeling the entire system with all the functionalities of the IP cores, only the network itself is modeled as communication between the cores is regarded to be the bottleneck in complex systems.

The approach for model creation is based on the studies for evaluating multistage interconnection networks (MINs) [4, 8, 15]. Generally, the main idea of those models is that a Markov chain represents the state of the first packet in a buffer. Another Markov chain represents the fill level of the corresponding buffer. The state transition probabilities are determined by calculating the sending and receiving probabilities of packets. This leads to a complex system of equations. Performance measures can be identified and computed solving the systems of equations. As performance measures the normalized throughput S_i at the inputs of the network, the normalized throughput S_o at the outputs of the network, and the mean delay time d_{tot} of the packets can be considered. The mean queue length $\overline{m}_{SE}$ of a specific buffer is also observable. For instance, this is useful to identify potential hot spots in the network under investigation. To solve the system of equations, fixed-point iteration techniques are usually applied [15] to obtain performance measures for the steady state of the network. This technique allows the investigation of the transient behavior, too, if particular constraints are fulfilled [14].

All these models have in common that they are only suitable for symmetric MINs. Asymmetric networks are not considered. Another aspect:

the models assume that packets are completely transferred from one stage to the next during a single clock cycle. A more realistic switching scheme like partial cut-through is not considered at the moment. This reduces model complexity but it is not feasible to obtain realistic performance measurements for on-chip networks.

To manage irregular networks and asymmetric traffic, all storage elements in the network have to be considered. This leads to a complex model: each pipeline register, buffer, even the ACK-registers have to be represented by at least one Markov chain describing this submodel which resulted from model decomposition. However, the development of the equations is simplified by automatic equation generating techniques [15]. The equations must only be set up once for each type of register and buffer. For a specific network configuration, the system of equations will be generated automatically. The complexity of the state space is manageable due to the limited number of cores on recent SoCs. Roughly 30 cores should be the upper limit for the next years.

The starting point of the model generation is a register transfer level description of the network. This description will be evaluated independently and translated into finite state diagrams for each buffer and register. Dependences between the state spaces are then identified. Next, the state diagrams are translated to discrete-time Markov chains, the identified dependences are transformed into equations of sending and receiving probabilities. This results in the state transition probabilities of the Markov chains. Because of the decomposition of the router functionality into several submodels, the resulting probability equations are simpler than in comparable models.

IP cores modeled as traffic sources provide the traffic for the network. Traffic analysis of real application has to be performed to describe traffic with stochastic distributions.

Three distributions for each traffic source have to be combined to reproduce the real system behavior. The distribution in time, the spatial distribution, and the multicast distribution of the traffic have to be considered.

The distribution in time determines the probability whether a packet is produced at the source. Multicast and spatial distributions determine which and how many destinations will be target of the packet. This influences the behavior of the routers as much as the chosen routing scheme. This superposition of different distribution is reflected by a routing matrix for each router in the network. The matrix denotes the probability that a packet from Input i is destined to Output j. For instance, in a symmetric network with uniform traffic distributions all elements in the matrix have the same value.

Proceeding in the above manner leads to a model where the mapping back to the real hardware becomes easy. With the probability of a certain state in the Markov chain, it is possible, for instance, to give a statement about the blocking probability in the input buffer of a specific router or the ratio of retransmittings on a specific link. A fine-grained analysis of the network in question becomes possible.

4 Possible Application Areas

In general, analysis of the performance of a specified on-chip network setup, usually calls for time-consuming simulations. With the presented architecture and the corresponding analytical model, one is now in a position to evaluate a large variety of different setups and topologies.

In interaction with models for the power consumption and area requirements, one is further able to optimize the network parameters for a given application. For instance, the NetChip framework [2] (consisting of SUNMAP [11] and × pipes) provides a complete synthesis flow for on-chip networks including topology selection. However, a variety of topologies has to be prepared for NetChip by hand. The presented performance model in cooperation with optimization algorithm could fill this gap to a fully automated framework. This is very helpful for the high-level synthesis community where large designs should automatically be generated from abstract descriptions.

With the introduction of dynamically reconfigurable devices such as FPGAs, dynamically reconfigurable network architectures [10] have been proposed, too. There, network topologies are modified during runtime. To identify optimal topologies for given traffic patterns fast analytical methods are needed to evaluate a large set of suitable configurations. In addition, the transient behavior of the network during reconfiguration is also a current research topic [9].

5 Conclusions

In this paper, an on-chip network architecture designed especially for embedded systems is presented. Furthermore, a corresponding stochastic model based on discrete-time Markov chains is proposed to enable fast and cycle accurate performance evaluation of the corresponding on-chip network. The stochastic model generation is still work-in-progress. In addition, a synthesizable implementation of the network architecture will be developed in a Hardware Description Language like VHDL or SystemC. The goal of this work is to present an analytical model, a simulation [16], and a realizable model of an on-chip network to improve SoC design.

Nevertheless, the proposed performance evaluation model of on-chip networks lacks of two important features. First, it is assumed that succeeding packets are mutually independent. No correlation between packets exists. It is not possible to emulate bursts of packets with this model. This leads to too optimistic results if the correlation of packet sequences is neglected. Extensions of the model are necessary to deal with the behavior of correlated packets.

Second, systems where real-time constraints have to be met, like guaranteed throughput or a maximum delay for certain messages, cannot properly be investigated with the proposed model. The randomness that is introduced in the arbiter of the routers does not allow priorities for certain packets. Nevertheless, this problem could be overcome by separation of best-effort traffic – covered with the model in this paper– and traffic with a guaranteed service (Quality of Service) character [12]. Separation could be realized by extending the network architecture with virtual channels: one channel for best-effort traffic and one or more for the traffic with real-time constraints.

References

1. Benini L, De Micheli G (2002) Networks on chips: A new SoC paradigm. Computer 35(1):70–78
2. Bertozzi D, Jalabert A, Murali S, Tamhankar R, Stergiou S, Benini L, De Micheli G (2005) NoC synthesis flow for customized domain specific multiprocessor Systems-on-Chip. IEEE Transactions on Parallel and Distributed Systems 16:113–129
3. Bolotin E, Cidon I, Ginosar R, Kolodny A (2004) Cost considerations in network on chip. The VLSI J, special issue on Network on Chip 38:19–42
4. Chan KS, Yeung KL, Chan SC (1997) A refined model for performance analysis of buffered banyan networks with and without priority control. In: Proceedings of the Global Telecommunications Conference Volume 3. Phoenix, AZ, USA, pp 1745–1750
5. Dally WJ, Towles B (2001) Route packets, not wires: on-chip interconnection networks. In: Proceedings of the 38th Conference on Design Automation. ACM Press, New York, pp 684–689
6. Hu J, Marculescu R (2004) Application-specific buffer space allocation for networks-on-chip router design. In: Proceedings of the International Conference on Computer Aided Design. San Jose, CA, USA, pp 354–361

7. Jalabert A, Murali S, Benini L, De Micheli G (2004) ×pipesCompiler: A tool for instantiating application specific networks on chip. In: Proceedings of the 41st Design, Automation and Test in Europe Conference and Exhibition. ACM Press, New York, pp 884–889
8. Jenq YC (1983) Performance analysis of a packet switch based on single-buffered Banyan network. IEEE J on Selected Areas in Communications SAC–1(6):1014–1021
9. Lüdtke D, Tutsch D, Walter A, Hommel G (2005) Improved performance of bidirectional multistage interconnection networks by reconfiguration. In: Proceedings of the 2005 Design, Analysis, and Simulation of Distributed Systems. San Diego, USA, SCS, pp 21–27
10. Majer M, Bobda C, Ahmadinia A, Teich J (2005) Packet routing in dynamically changing networks on chip. In: Proceedings of the 19th IEEE International Parallel and Distributed Processing Symposium. Washington, DC, USA, IEEE Computer Society, 154.2 vol 4
11. Murali S, De Micheli G (2004) SUNMAP: A tool for automatic topology selection and generation for NoCs, In: Proceedings of the 41st Conference on Design Automation and Exhibition. ACM Press, New York, pp 914-919
12. Rijpkema E, Goossens K, Rădulescu A, Dielissen J, van Meerbergen J, Wielage P, Waterlander E (2003) Trade-offs in the design of a router with both guaranteed and best-effort services for networks on chip. IEE Proceedings Computers & Digital Techniques. 150(5), pp 294–302
13. Stergiou S, Angiolini F, Carta S, Raffo L, Bertozzi D, De Micheli G (2005) × pipes lite: A synthesis oriented design library for networks on chips. In: Proceedings of the Design, Automation and Test in Europe Conference and Exhibition. Munich, Germany, pp 1188–1193
14. Tutsch D (1999) Performance analysis of transient network behavior in case of packet multicasting. In: Proceedings of the 11th European Simulation Symposium. Erlangen, Germany, pp 630–634
15. Tutsch D, Hommel G (2002) Generating systems of equations for performance evaluation of multistage interconnection networks. J of Parallel and Distributed Computing 62(2):228–240
16. Tutsch D, Lüdtke D, Walter A, Kühm M (2005) CINSim – A component-based interconnection network simulator for modeling dynamic reconfiguration. In: Proceedings of the 12th International Conference on Analytical and Stochastic Modelling Techniques and Applications. Riga, pp 132–137
17. Wang H-S, Zhu X, Peh L-S, Malik S (2002) Orion: A power-performance simulator for interconnection networks. In: Proceedings of the 35th annual ACM/IEEE international symposium on Microarchitecture. IEEE Computer Society Press, Los Alamitos, CA, USA, pp 294–305
18. Warnakulasuriya S, Pinkston TM (2002) Characterization of deadlocks in irregular networks. J of reParallel and Distributed Computing 62(1):61–84

Simulation-Based Testing of Embedded Software in Space Applications

Sergio Montenegro[1], Stefan Jähnichen[2], and Olaf Maibaum[2]

[1] Fraunhofer FIRST
[2] Computer Architecture and Software Technology
Technische Universität Berlin, Germany/Fraunhofer FIRST
[3] Simulation and Software Technology
Deutsches Zentrum für Luft- und Raumfahrt e.V., Germany
{Stefan.jaehnichen,Sergio}@first.fhg.de, Olaf.Maibaum@dlr.de

Summary. This paper deals with the software-in-the-loop test approach being developed by the consortium project SiLEST (DLR, TU-Berlin, IAV, FhG FIRST, Webdynamix). We present a layer structure of the control loop that allows components of the environment simulation to be used for hardware-in-the-loop and software-in-the-loop testing of embedded systems software. The approach is specifically designed to test software behaviour in disturbed operating conditions, such as in a harsh environment, for example. In space applications, intensive radiation can corrupt computations and stored data. In addition, electronic devices such as sensors age much faster than on earth so that changed sensor deviations must be expected. Much the same is true of numerous other embedded systems, e.g. in automotive applications. Here, too, the electronic components are exposed to extreme conditions (temperature) and are subject to ageing processes.

1 Introduction

The development work was motivated by difficulties encountered in testing the BIRD satellite. BIRD is a technology demonstrator and a space fire alarm (see Figure 1).

The BIRD microsatellite mission demonstrates the technical and programmatic feasibility of combining ambitious science and new – not yet space-proven – advanced technologies under fixed budget constraints. Demonstrating new microsatellite technologies is a key objective of the BIRD mission. The technology experiments demonstrate the limitations and advantages of the newly developed components and technologies. The BIRD microsatellite (mass = 92 kg) was launched with the Indian PSLV-C3 from Shar on 22 October 2001 into a sun-synchronous circular orbit at an altitude of approx. 568 km. BIRD orbits the earth every 90 minutes, scanning a 300 km-wide band. It is capable of detecting, analyzing and reporting fires of 8 sq m in size upwards. BIRD's fire-detecting capabilities

Günter Hommel and Sheng Huanye (eds.), Embedded Systems Modeling – Technology, and Applications, 73–82.

exceed the most optimistic expectations. Since its launch, BIRD has functioned with high reliability, despite some radiation-related problems and hardware failures.

Fig. 1. BIRD in orbit

Most functions in BIRD are software-controlled using the real-time operating system and middleware BOSS. BOSS was designed for space applications, in particular for dependability, safety and simplicity. Our aim was (and is) to attain the greatest possible dependability of embedded systems by reducing development errors (through simplicity) and handling runtime anomalies (by fault-tolerance support). The principles underlying the construction of BOSS and its middleware were: design and build an irreducibly complex system using modern framework technology for the underlying operating system and component technology for the middleware and its applications. The results are very promising. BOSS has been in continuous use in space (BIRD satellite) and in medical devices for a number of years now. Even complex functionality can be implemented very easily using BOSS.

Ensuring BIRD's correct functioning was a very difficult task, not only because the functionality is largely implemented in software but also because BIRD makes use of numerous new technologies, making the test phase particularly important and difficult. Some functions are not easy to test on the earth: for instance, attitude control requires zero gravity and the solar sensors need extremely high light intensity. For such cases, a highly complex testbed was built (see figure 2). For future missions, it is planned to reduce this complexity, though not at the expense of reliability. That is why we are building a virtual simulation environment for satellites, which can be used equally well for other embedded systems.

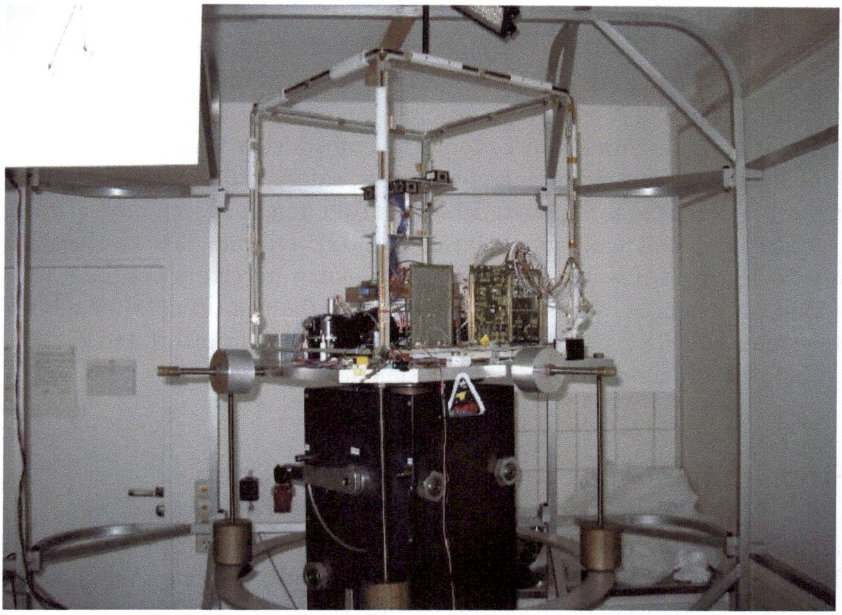

Fig. 2. Testbed

2 Testing Adaptive Systems

The environmental conditions in many embedded systems are non-optimal. In space applications, radiation is a problem, while earth-bound systems, e.g. in the automotive sector, are subjected to cycles of heat and cold, moisture and corrosion, vibrations, fouling and mechanical wear, which cause sensors and actuators to age. The software-implemented control system must take this into account by using adaptive mechanisms to adapt to gradually changing conditions. Such adaptive mechanisms may, for example, involve adapting filters for sensor data or reconfiguring redundant sensor networks, which will be essential for future X-by-wire systems and are customary in space applications to extend the useful life of on-board systems.

Given the variety of ageing processes and failure events affecting sensors and actuators and the high complexity of sensor networks, the testing of these adaptive software mechanisms represents a considerable challenge. SiLEST will incorporate a "software-in-the-loop" (SiL) simulation to test software behaviour in an aged system for the automotive and aerospace application domains. “Software-in-the-loop” testing means that testing of the real software – including all restrictions with regard to resources – is carried out in a simulated environment or using experimental hardware.

Furthermore, the process of developing embedded software systems in a technical environment with high correctness, safety and robustness requirements often accounts for approx. 50-80% of the total development effort in the verification and validation phases [RB02]. Despite this considerable effort for testing software, potentially fatal software errors continue to be common occurrences. By seeking to improve the test process for embedded systems, SiLEST promises to reduce development effort and increase the reliability of embedded software systems.

3 From Real Devices to Software-in-the-Loop

In addition to classical functional and module testing, embedded systems software must also be tested in the complete controller loop including target hardware and target devices (see Figure 3).

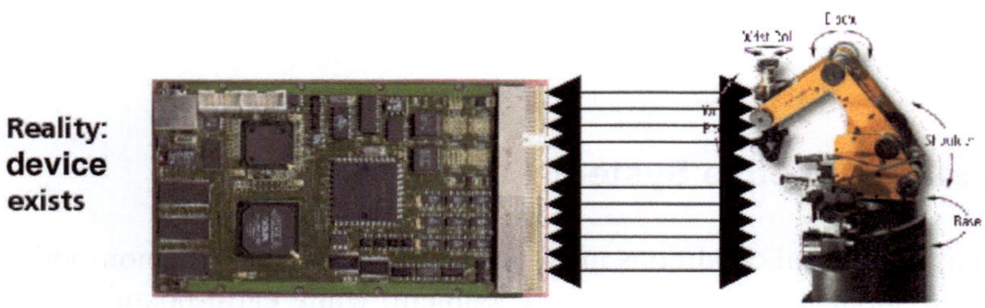

Fig. 3. Control software in a real device

Embedded systems software is an integral part of this loop and cannot be considered separately from it. Every output of the software in the loop triggers feedback on the sensor data. This must be taken into account when testing embedded systems software. The conventional approach to testing embedded systems using simulation is the HiL (Hardware-in-the-Loop) test (see Figure 4)

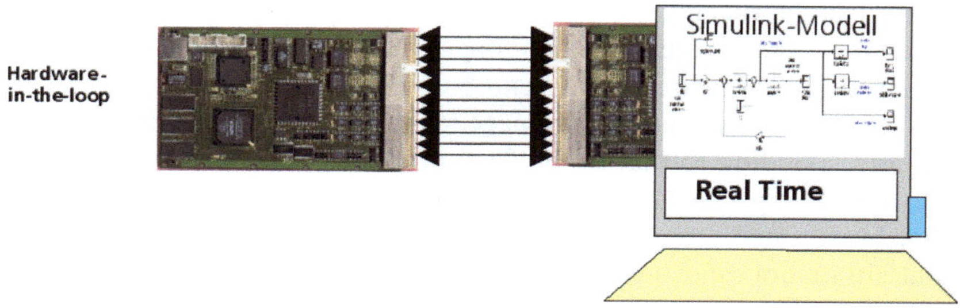

Fig. 4. Hardware-in-the-loop

This approach uses laboratory prototypes of the technical system under test or simulations. Testing using a laboratory prototype is highly cost-intensive and can only be carried out at a very late stage in the development process because the relevant system hardware has to be available for testing. Simulation testing involves simulating the environment, the sensors and the actuators of the system being tested. The sensors' and actuators' simulation is coupled via the processor board's hardware interfaces and special VME or PCI interface cards (a complex business).

The HiL test approach has a number of drawbacks. The hardware needed for the test setup is not normally available until late on in the development process, and often then there is only a single specimen of the Hardware Simulator, which is very expensive. This makes the test setup cost-intensive and laboratory-bound. The coupling of the simulation and the embedded system makes it difficult to keep a consistent check on the state of the software and the environment. This, in turn, makes it hard to diagnose errors in the embedded system.

Our alternative to HiL testing is to use a SiL (Software-in-the-Loop) test approach. With this approach, the system under test and the environment simulation are coupled via a single communication connection (e.g. Ethernet) (see Figure 5).

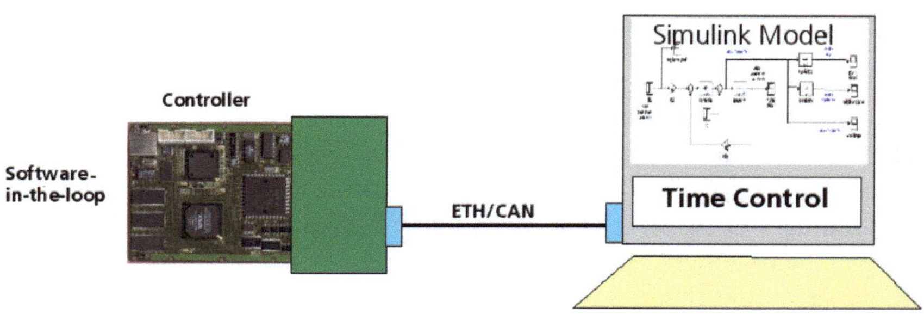

Fig. 5. Software-in-the-loop

The SiL test approach enables the in-circuit test harness to be dispensed with and means that the testing is no longer laboratory-bound. And with no need for special hardware, testing can begin early on in the development process. The only essential requirement for testing is a controller approximately corresponding to that in the final product in terms of temporal behaviour and storage capacity and a simulation computer. By dispensing with hardware interfaces, SiL testing allows close coupling of the environment simulation and the system under test. This coupling method enables a consistent state of the simulation and the software under test to be achieved, which can be used for error diagnosis and as an initial state for further test runs. Figure 6 shows the simulation possibilities for different test approaches.

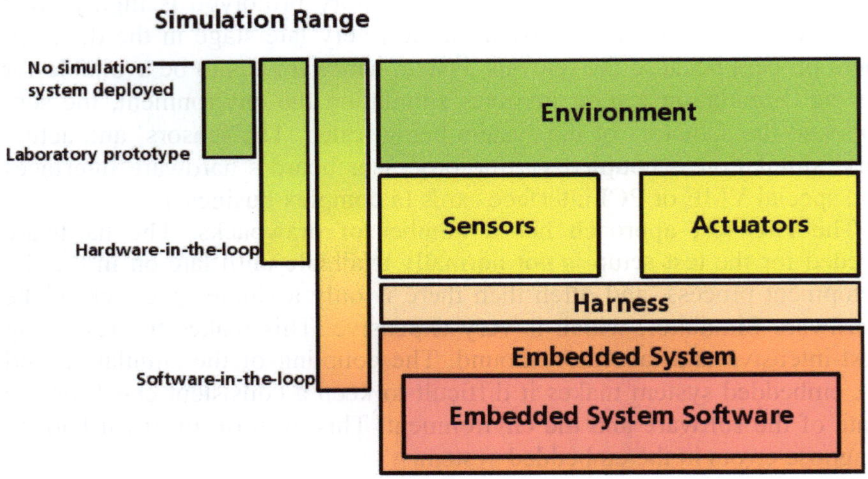

Fig. 6. Test approaches

4 Our SiL Approach

We aim to replace the real environment of the embedded software by a simulation, without the embedded software being aware of this (see Figure 7). In a real system, the operating system executes I/O accesses using the corresponding I/O drivers. Our SiL approach dispenses with I/O devices and drivers altogether. Instead, all I/O accesses are converted to UDP messages (a simple Internet protocol), which are sent by Ethernet (ETH) to a remote simulator. The simulator computes the environment reaction and returns the simulated environment conditions, also using UDP messages too.

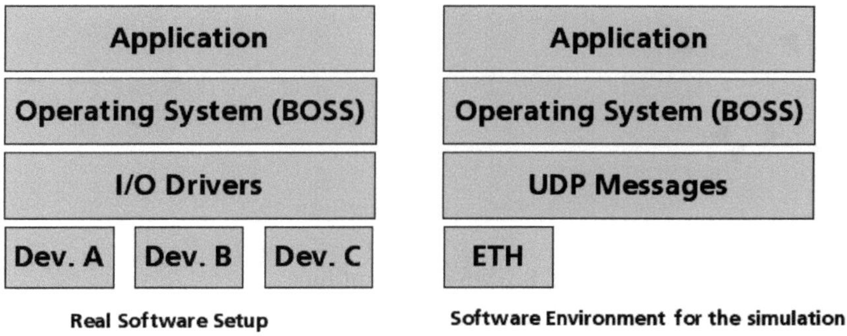

Fig. 7. Embedded software for the SiL test

The biggest problem faced here is the system's time behaviour. HiL simulations must run in real time, which makes the simulator highly complex and expensive. Our SiL simulation does not run in real time, but the embedded software is not aware of any time delays, caused by the UDP communication and simulation steps. This is the challenge confronting us.

To the embedded software, the system appears to run in real time. To achieve this, the operating system (BOSS) was extended by the addition of a virtual time manager, which can freeze the time and all software activity while UPD communication and synchronization is under way. This means that the virtual time is not continuous. The system has two notions of time: virtual/simulation time and real time (see Figure 8). The virtual time of the embedded software has to be synchronized with the simulation time (they should be the same) in such a way that to the embedded software the simulation time appears to be real and continuous. By contrast, the (real) real time is that perceived by an external observer. Since the simulation runs slower than the real world, the time of the embedded system runs faster than the real time. To keep them synchronized, the virtual time has to be stopped every time the embedded software tries to access the external (simulated) world.

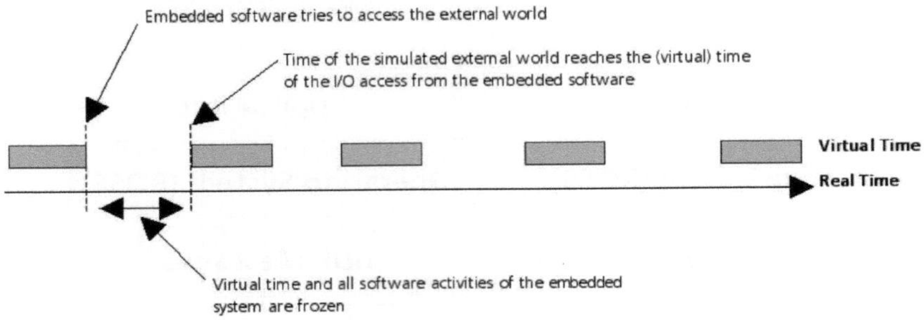

Fig. 8. Two notions of time

To synchronize the virtual time and data in the embedded software with the simulation time and data in the simulator, we introduced a communication protocol which combines data transfer and time synchronizations, as shown in Figure 9.

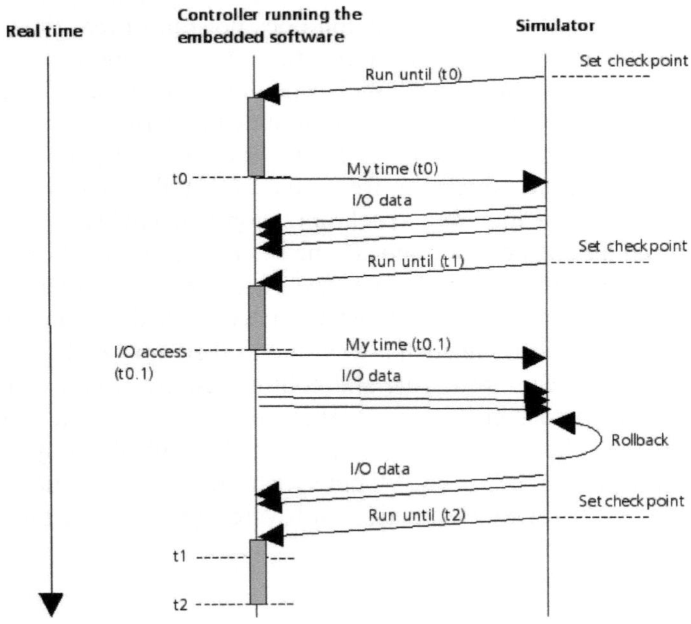

Fig. 9. SiLEST synchronization protocol

The simulator begins the the simulation and continues until a time point (let us call it T0) at which it detects an asynchronous signal (data) going to the embedded controller. This might be, say, certain conditions generating an interrupt in the controller. The simulator thus knows when it will interrupt the controller, but it does not know when the controller will access data from I/O devices. The simulator allows the controller to run until a determined time point (message "Run Until(T0)" in Figure 9). The controller runs until T0 at the most, at which point the virtual time manager stops the time and all other software activities in the embedded controller. It reports the time to the simulator and waits for further instructions. At this point, the simulator transfers data to the controller. The simulator then gives the clearance to continue running until T1 (message "Run Until (T1)"). If, before this time point is reached, the embedded software attempts to access an I/O device (this is mostly the case), the virtual time manager again stops the virtual time and all other software activities, and then reports the time and the I/O access data to the simulator. The simulator receives the data and eventually it has to perform a rollback to the last checkpoint and resimulate until T0.1, which is before T1. At this point, a bidirectional data transfer is carried out to synchronize data and time. After these synchronizations, the simulator gives the clearance to continue until the next asynchronous signal – T2 in Figure 9 – which may still be the same as T1.

According to this protocol, there are two reasons for stopping the virtual time: I/O access and the reaching of the maximal simulation step given by the simulator.

5 Conclusions and Outlook

The SiL test approach presented here dispenses with the need for specialized hardware, making it suitable for deployment early on in the development process. Also, its use of synchronization between the controller and the simulation makes it possible to take a more precise look at the state of the software than is possible by an HiL test environment. Debugging potential in particular is enhanced by the ability to execute the software and simulation in steps. The chosen approach also makes it possible to obtain consistent states of the environment and the software under test. The well-defined layer structure of the simulation environment allows a library of simulation modules to be compiled, which can be used for HiL/SiL tests and, at the same time, is suitable for cross-project application. This enables costs to be cut during test preparation.

The SiL test's practical potential and limitations will only become evident towards the end of the SiLEST project, when the test process is tried out on software from the automotive and aerospace application domains. At present, it can safely be stated that real-life and HiL testing cannot be completely replaced by SiL testing because previous experience has shown that simulation and reality always differ. However, much of the environment simulation used in SiL testing will also be suitable for use in HiL tests and the flexibility offered by a totally software-implemented test environment will outweigh the drawback of the additional effort required by early error detection and the improved debugging options.

6 Recommended further literature and references

Briess K, Baerwald W, Gill E, Halle W, Kayal H, Montenbruck O, Montenegro S (2003) Technology demonstration by the bird-mission, ISBN 3-89685-569-7.

Kumar K, Goswami, Ravishankar K (1993) Simulation of Software Behaviour Under Hardware Faults. Proc. of the 23rd Int. Symp. on Fault Tolerant Computing. 1993. pp. 218-227.

Poncet, JC (2000) Using Simulation to Design Real Time Applications. In Simulation in Industry 2000, 12th European Simulation Symposium 2000. Ed. D.P.F. Möller. Hamburg, Sep. 2000. pp. 43-47.

Raguideau J, Schoen D, Henry J (1994) an Event-Driven Simulation Tool for Testing Software. In 5th Int. Symp. on Software Reliability Engineering. 1994. pp. 259-263.

SILEST Web presentation: www.silest.de

Zimbelman D, Anderson M, Correl T, Schnurr R, Fennel M (1996) The Attitude Control System Test-Bed for SWAS and future SMEX Missions. In Guidance and Control. 1996. V. 92. pp. 145-163.

Evolving Specifications for Embedded Systems in the Automotive Domain

André Metzner and Peter Pepper *

Language Design and Compiler Construction
Technische Universität Berlin, Germany
{ame,pepper}@cs.tu-berlin.de

Summary. In many respects a modern luxury car resembles a complex software system on wheels, containing over 70 embedded systems, many of them interconnected using domain-specific networks. In order to cope with the still-increasing complexity, there is a strong need to specify system characteristics formally, even for security-insensitive subsystems. In this paper we present a case study of a basic car entertainment system, specified in part by using *evolving specifications* (especs). Especs have a thorough mathematical foundation in category theory, but do not require deep categorical knowledge to be deployed. We use visualizations to guide the reader through our espec design.

1 Introduction

Modern automobiles contain complex software systems consisting of dozens of embedded systems, organized within various domain-specific heterogeneous sub-networks. Within the telematics domain on its own we find AM/FM radios, CD/DVD/MP-3 players, TV sets, GPS navigation systems, cellular communication facilities, etc. The ever-increasing complexity of these devices and their interactions with each other has made reliability and user experience major factors of concern. Thus a common requirement in the development cycle is the formal specification of functional and operational behaviors.

Evolving specifications (especs), developed by Duško Pavlović and Douglas R. Smith at Kestrel Institute [4, 5, 6, 7], are related to Gurevich's evolving algebras (abstract state machines) [2], but take care to embed themselves into the rich mathematical framework of category theory. Categorical notions like refinement morphisms and colimits of diagrams provide clean semantics for top-down development and specification composition. Refinement can potentially go down to the implementation level, making especs promising candidates for the development of *specification-carrying software.*

* This work was sponsored in part by funding from DaimlerChrysler.

Günter Hommel and Sheng Huanye (eds.), Embedded Systems Modeling – Technology, and Applications, 83–92.

Category theory has the reputation of being a very abstract branch of mathematics. Many potential users of especs are likely to disregard them just because of their categorical background, overlooking the fact that the mechanisms utilized by especs are basic enough to be automated, as proven for example by the Epoxi system [6]. In an attempt to make especs more widely known, we present a case study of a basic telematics system that we expand piece by piece over the course of Sects. 3 to 5 after a short review of basic espec concepts in Sect. 2. Section 6 concludes this paper.

2 Evolving Specifications

A template for an espec looks as follows:[2]

```
ESPEC ≪name≫(≪env espec≫          // parameterization by environment(s)
             VIA ≪parameter morphism≫...)
  IMPORT ≪espec≫...               // modularization
  // algebraic parts
  SORT ≪sort≫...                  // global specification of sorts,
  OP ≪operation≫...               //    operations, variables,
  AXM ≪axiom≫...                  //    and properties
  // coalgebraic parts
  INITIAL-MODE ≪mode≫             // the component's start-up mode
  MODES ≪mode_1≫,≪mode_2≫,...     // modes without local structure
  MODE ≪mode≫                     // mode with local structure
    ≪specification extension≫...
  ...
  TRANS ≪trans≫: ≪mode_src≫ → ≪mode_dst≫
    ≪guard≫ ⊢ ≪effect≫            // conditional transition between modes
  ...                             // (empty guard or effect is written as ⊤)
END-ESPEC ≪name≫
```

Ignoring the parameterization for a moment, we find *modularization* support and identify sorts, operations and axioms as components of an *algebraic specification* (spec) [1] that describes *functional* behavior and provides a global *structure* underlying the espec. Modes (also known as state descriptions) and transitions describe *operational* behavior and form a kind of state machine.

We prefer the term "mode" over "state" because it emphasizes that the ressource modeled by the espec is usually quite active in a mode. Part of the coalgebraic view on especs is the assumption that we do not know the system in complete detail, but rather *observe* certain characteristics. Thus a mode describes an *abstract state* rather than a concrete state. This abstract state denotes a set of models of an *extension of the global spec* defined by

[2] Algebraic and coalgebraic parts are traditionally marked by SPEC and PROG keywords, but we omit these (redundant) labels for brevity.

the local mode description. A mode thus comes with a (theorem-preserving) morphism within the category of specs with the global spec as domain. The evolving thing about especs is the change of the currently active spec by a transition. The transition is associated with a *translation* (morphism) from the destination mode spec back to the source mode spec, similar to the reasoning in Hoare logic.

The *environment parameters* are especs reflecting the view of a ressource on its environment [5]. As interface descriptions they permit to hide inner details of the ressource. We allow for multiple parameters in order to enable the designer to separate different aspects. Looking ahead to Figs. 3 and 4, one finds that each parameter espec is connected to its ressource espec by a *parameter morphism*. The mixed algebraic and coalgebraic nature of especs implies that a morphism from A to B maps the global spec in forward direction, but the modes and transitions in backward direction, i.e. from B to A. The connection between an environment and a ressource implementing this environment is given by a *refinement morphism*. The construction of a whole system is performed by taking the colimit of a diagram of ressources and environments in the category of especs.

Due to space restrictions we cannot elaborate on the categorical foundations sketched above, but are confident that the involved concepts will become more clear by studying our example in the following sections. For further details about especs please refer to [4, 6, 7].

3 A Lone CD Player

We begin by connecting a CD player to an amplifier. Although there is no immediate need to mediate access to the amplifier, we anticipate the later addition of other components and place a controller in-between as shown in Fig. 1. We mention on a side note that entertainment devices in modern cars are not wired in a peer-to-peer fashion, but utilize communication bus systems like the Media Oriented Systems Transport (MOST) [3]. The MOST bus is quite sophisticated though cost-effective and able to transfer control data as well as to stream multimedia data on multiple channels simultaneously.

The minimum functionality the CD player has to provide is to react to "play" and "stop" button presses by the user and to tell the rest of the system about it. In addition, we expect the CD player to also act on equivalent commands received on the control channel. Within the espec framework, communication between components is usually modeled in a rather indirect way by shared variables: a boolean variable tells about a pending transmission and another variable contains the data. We prefer to hide these mechanisms in favor of a message-passing style which makes communication explicit. The special primitives `SND` ≪msg≫ and `RCV` ≪msg≫ act on typed messages for which we introduce suitable data types below. The message constructors are

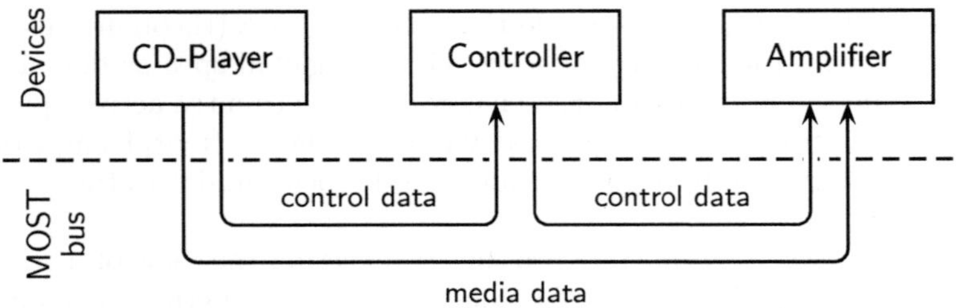

Fig. 1. A basic car entertainment system with logical interconnections of devices. Physically all devices are connected to a common bus system.

also used for message decomposition at reception time in a way resembling pattern matching as known from functional languages.

```
// supporting data types
TYPE Channel  == None | Channel(Nat)              // simple channel type
TYPE DeviceId == None | CD | Radio | ...          // device identification
// message data types
TYPE Play     == Play(DeviceId)                   // controller → device
TYPE Stream   == Stream(DeviceId, Channel)        // device → controller
TYPE Stop     == Stop(DeviceId)                   // device ↔ controller
TYPE PlayBack == PlayBackChannel(Channel)         // controller → amplif.
TYPE StopChannel == StopChannel(Channel)          // controller → amplif.
```

To allow for later reuse, we introduce a generic `PlayableDevice` parameterized by a `DeviceId` that is supposed to identify the particular device, e.g. the CD player espec simply instantiates this espec via IMPORT `PlayableDevice[CD]`. The device behavior is depicted graphically in the left part of Fig. 2. We use some syntactic sugar in the following espec as a shorthand: a transition TRANS t with ROLE a and ROLE b actually denotes two transitions t_a and t_b with common domain and codomain. For clarity, visualizations show only a single (multi-)transition t_* instead of parallel transitions.

```
ESPEC PlayableDevice[self: DeviceId]          // instance parameter
        (PlayableDevice-Env VIA               // environment parameter
          play ↦ stream_snd, play' ↦ stream_rcv,
          stop_snd ↦ stop_rcv, stop_rcv ↦ stop_snd)
  IMPORT ChannelOps                     // for channel allocation handling
  VAR channel: Channel                  // variable ≡ nullary operation
  MODES Stand-By, Playing-Requested, Playing
  INITIAL-MODE Stand-By
  TRANS play: Stand-By → Playing-Requested
    ⟨User⟩ ∨ (RCV Play(dev^) ∧ dev^ = self) ⊢ ⊤
  TRANS play': Playing-Requested → Playing
    ⊤ ⊢ channel := allocChannel(); SND Stream(self, channel)
```

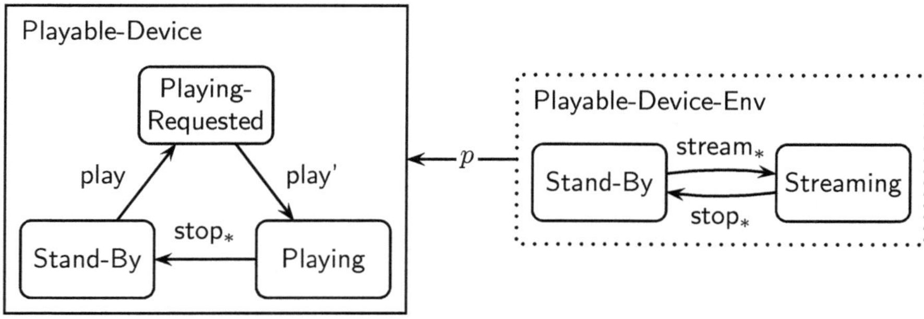

Fig. 2. A device with basic playing capability and its environment. They are related to each other by a parameter morphism.

```
  TRANS stop: Playing → Stand-By
    ROLE snd: ⟨User⟩ ⊢ deallocChannel(channel); SND Stop(self)
    ROLE rcv: RCV Stop(dev̂) ∧ dev̂ = self
                ⊢ deallocChannel(channel)
END-ESPEC PlayableDevice
```

The component begins its operation in mode `Stand-By` and switches to mode `Playing-Requested` either on user interaction or by reception of a `Play(...)` message with a `DeviceId` matching its own; $\widehat{\texttt{dev}}$ denotes a local variable to be bound by pattern matching. After allocation of a transmission channel, a streaming request is sent. The actual media data transmission happens during the `Playing` mode, but that is an aspect not reflected by the espec since we are only concerned with control data exchange. There are two ways to cause the device to stop playing: by user interaction (transition $\texttt{stop}_{\texttt{snd}}$; a `Stop(...)` message is *sent*) or by command (transition $\texttt{stop}_{\texttt{rcv}}$; a `Stop(...)` message is *received*).

To increase design flexibility, devices are not directly bound to other devices. Expectations and requirements about a device's surrounding are specified in environment especs that essentially mirror the communication behavior: `SND` operations of a device correspond to `RCV` operations in its environment and vice versa. Parameter morphisms are given by transition mappings in device especs; mode mappings are induced implicitly. In a similar way, but not inverting `SND`/`RCV`, devices implementing environment expectations are connected with environment especs by refinement morphisms.

```
ESPEC PlayableDevice-Env
  MODES Stand-By, Streaming
  INITIAL-MODE Stand-By
  TRANS stream: Stand-By → Streaming
    ROLE rcv: RCV Stream(dev, chan) ⊢ ⊤
    ROLE snd: ⊤ ⊢ SND Play(dev)
  TRANS stop: Streaming → Stand-By
```

```
    ROLE rcv: RCV Stop(dev) ⊢ ⊤
    ROLE snd: ⊤ ⊢ SND Stop(dev)
END-ESPEC PlayableDevice-Env
```

We demand that a parameter morphism ensures the desired SND/RCV correspondence. However, as a side effect this requirement effectively inhibits the use of multiple message transmissions within a single transition. To see why, consider the transitions `play` and `play'` from above. These are the only transitions leaving their domain modes and `play` has no effect except to change the mode. It seems natural to combine the two transitions into one, getting rid of mode `Playing-Requested` along the way:

```
  TRANS play: Stand-By → Playing
    ⟨User⟩ ∨ (RCV Play(dev^) ∧ dev^ = self)
      ⊢ channel := allocChannel(); SND Stream(self, channel)
```

However, the problem is that the parameter morphism cannot map the new `play` transition simultaneously to $\mathtt{stream}_{\mathtt{snd}}$ and $\mathtt{stream}_{\mathtt{rcv}}$. While the former separation may be a bit cumbersome, it is meaningful. Fortunately it is straightforward to automatically separate a "combined" transition into a sequence of transitions and to introduce appropriate intermediate modes. Therefore we allow the more compact way of expression and refer to an implicit RCV helper transition induced by transition t as $t^{\triangleleft}$ and to a SND helper transition as $t^{\triangleright}$ (or $t^{\triangleright_1}, t^{\triangleright_2}, \ldots$, if there is more than one SND).[3] The triangles hint to the left and right sides of the guard sign.

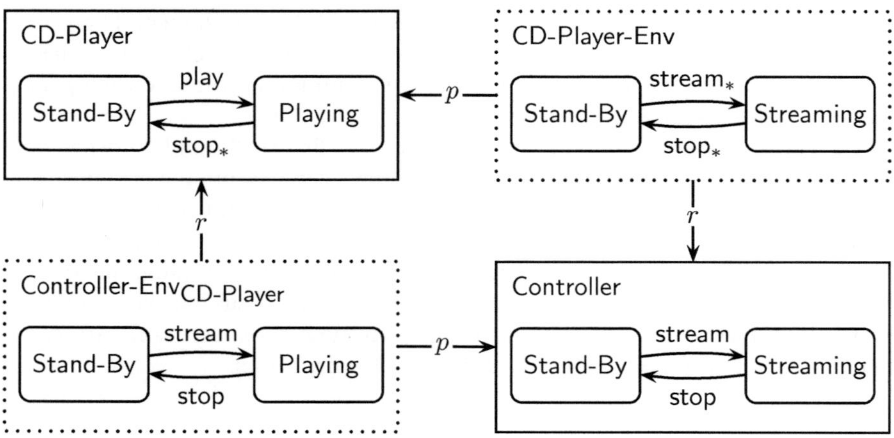

Fig. 3. The CD player refines the environment of the controller and vice versa. Refinement morphisms are marked by r, parameter morphisms by p.

[3] Multiple RCV's in a single guard are rare and should better be treated explicitly. In combination with (auto-)generated transitions we allow only for a single RCV.

Figure 3 illustrates that the controller component serves as implementation of the CD player's environment espec with a refinement morphism `CD-Player-Env` $\xrightarrow{r}$ `Controller` consisting of $\mathtt{stream}^{\triangleleft} \mapsto \mathtt{stream}_{\mathtt{rcv}}$ and $\mathtt{stop}^{\triangleleft} \mapsto \mathtt{stop}_{\mathtt{rcv}}$. However, not apparent from this figure is the interaction of the controller with the amplifier. We choose to separate this aspect into another environment (see Fig. 4). The amplifier needs to be instructed about the channel to stream and acts on `PlayBackChannel(...)` and `StopChannel(...)` messages emitted by the controller during $\mathtt{stream}^{\triangleright}$ and $\mathtt{stop}^{\triangleright}$ transitions. The controller tracks the channel number internally.

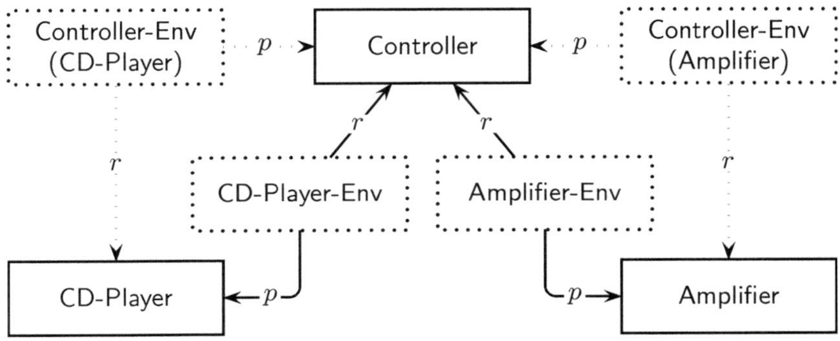

Fig. 4. The controller interacts with all components, but CD player and amplifier do not know each other for the purposes of control message exchange.

To build a whole system means to compose the components, i.e. to take the colimit of the diagram from Fig. 4 in the category of especs. At this stage the glueing provided by the morphisms becomes essential. Please refer to [6] for more details about colimit calculation of espec diagrams.

4 Adding a Radio

On our way to a more interesting system, we instantiate the `PlayableDevice` once again to incorporate a simple radio (see Fig. 5). This addition affects the controller that now has to mediate amplifier access, reflected by a new transition `switch` in mode `Streaming` (see Fig. 6).

$$
\begin{array}{l}
\mathsf{TRANS}\ \mathtt{switch:\ Streaming} \rightarrow \mathtt{Streaming} \\
\quad \mathsf{RCV}\ \mathtt{Stream}(\widehat{\mathtt{dev}},\ \widehat{\mathtt{chan}}) \vdash \mathsf{SND}_1\ \mathtt{StopChannel(channel);} \\
\qquad\qquad\qquad\qquad\qquad\quad \mathsf{SND}_2\ \mathtt{Stop(streamDev);} \\
\qquad\qquad\qquad\qquad\qquad\quad \mathtt{streamDev} := \widehat{\mathtt{dev}};\ \mathtt{channel} := \widehat{\mathtt{chan}}; \\
\qquad\qquad\qquad\qquad\qquad\quad \mathsf{SND}_3\ \mathtt{PlayBackChannel(channel)}
\end{array}
$$

The controller stops the amplifier, stops the currently active device, remembers information about the new device/channel for later use, and starts

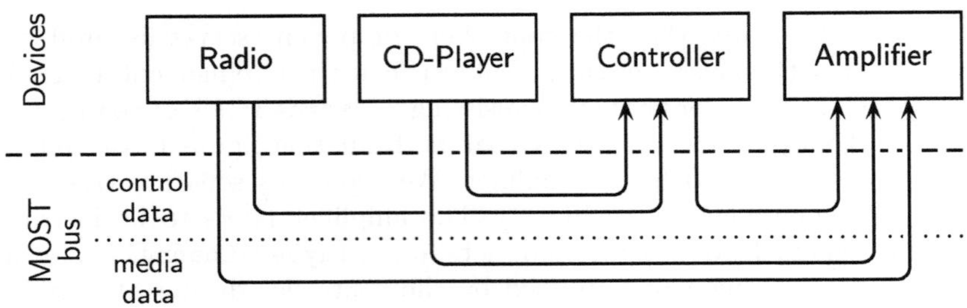

Fig. 5. The car entertainment system from Fig. 1 augmented with a radio.

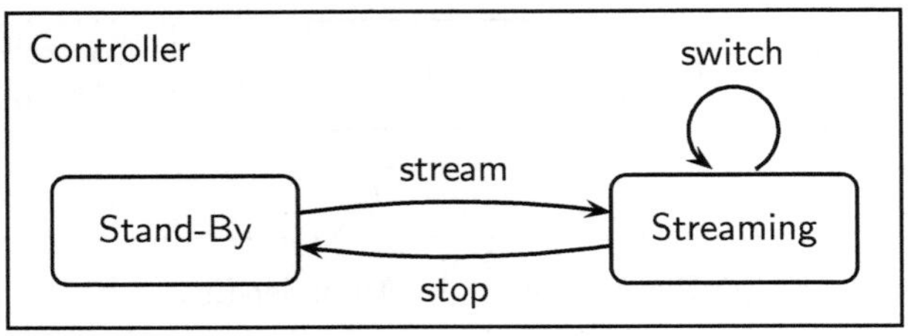

Fig. 6. The controller from Fig. 3 augmented with switching capabilities.

the amplifier again. The controller's environment is slightly affected by the change, because controlled devices must be prepared to receive `Stop(...)` messages which was not strictly necessary before (but the `PlayableDevice` already provides that feature).

For brevity, we have simplified the especs in our running example by omitting axioms that would be specified in practice. For example, it would be sensible to ensure that the `channel` variable is only non-null when there is a sending device and that this is indeed the case in mode `Streaming` before we try to stop this device.

5 Handling Traffic Announcements

The radio of the previous section lacks an important feature of real car radios: to announce traffic informations that are received by an internal second tuner. We introduce a new mode `Announcing-Traffic-Info`, reachable from both `Stand-By` and `Playing`. Upon entering and leaving this mode new messages `StreamPrio(DeviceId, Channel)` and `StopPrio(DeviceId)` are emitted.

A traffic info message should be played even when the radio is not the currently active device. However, after playing the traffic info, the system should resume to do what it did before; this behavior requires support from the controller. Figure 7 shows the controller augmented with, among other things, the new mode `Priority-Streaming`, active during traffic info announcements.

In general, there could be additional sources of priority informations, e.g. a navigation system, but we do not model the interruption of a priority

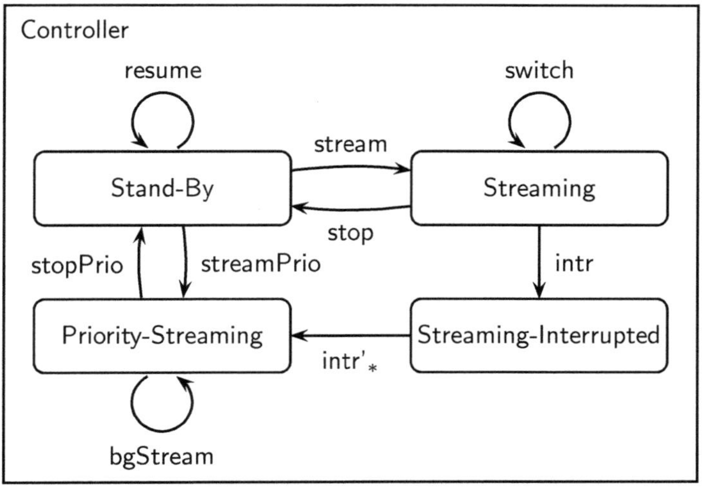

Fig. 7. The controller component supporting priority message handling.

source by another one. However, we do handle the interruption of a normal source by a priority source via the transitions **intr** and **intr'**$_*$. The mode **Streaming-Interrupted** is not strictly necessary, but conveniently factors out pattern matching and variable assignment:

TRANS intr: Streaming $\rightarrow$ Streaming-Interrupted
  RCV StreamPrio($\widehat{\texttt{dev}}$, $\widehat{\texttt{chan}}$)
    $\vdash$ streamPrioDev := $\widehat{\texttt{dev}}$; prioChannel := $\widehat{\texttt{chan}}$;

Depending on the identity of the priority device, the device currently playing and the channels to use, we differentiate three cases:

TRANS intr': Streaming-Interrupted $\rightarrow$ Priority-Streaming
  ROLE 1: streamPrioDev $\neq$ streamDev
    $\vdash$ SND_1 Stop(streamDev);
      SND_2 StopChannel(channel); channel := None
      SND_3 PlayBackChannel(prioChannel)
  ROLE 2: streamPrioDev = streamDev $\wedge$ prioChannel $\neq$ channel
    $\vdash$ SND_1 StopChannel(channel); channel := None
      SND_2 PlayBackChannel(prioChannel)
  ROLE 3: streamPrioDev = streamDev $\wedge$ prioChannel = channel
    $\vdash$ channel := None

If a device switch is requested during priority playing, we immediately stop this device, but remember the changed device for later:

TRANS bgStream: Priority-Streaming $\rightarrow$ Priority-Streaming
  RCV Stream($\widehat{\texttt{dev}}$, $\widehat{\texttt{chan}}$) $\vdash$ streamDev := $\widehat{\texttt{dev}}$;
    SND Stop(streamDev)

When priority playing is over, mode `Stand-By` is reached. If originally there was a normal device playing that got interrupted, the variable `streamDev` still refers to it and we instruct the device to start playing. It will then issue a `Stream(...)` message, subsequently causing a `stream` transition.

```
TRANS resume: Stand-By → Stand-By
  streamDev ≠ None ⊢ SND Play(streamDev); streamDev := None
```

6 Conclusions

Evolving specifications can be deployed in quite different contexts. We have presented a case study in the automotive telematics domain to illustrate several key concepts while skipping some technical aspects. Especs together with morphisms between them form a category, offering access to a rich framework of methods. Parameter morphisms associate ressource especs with environment especs acting as interface descriptions; refinement morphisms connect implementation especs to environments; colimits of espec diagrams describe system composition. Among the topics for future research are automation support and the modeling of advanced communication mechanisms.

Acknowledgement. We thank or colleagues from the ÜBB group at the TU Berlin for valuable comments. Particular thanks go to Duško Pavlović and Douglas R. Smith for many elucidating discussions about especs.

References

[1] Ehrig H, Mahr B (1985, 1990) Fundamentals of algebraic specification (vol 1 and 2). EATCS Monographs on Theoretical Computer Science, vol 6, 21. Springer
[2] Gurevich Y (1994) Evolving algebras 1993: Lipari Guide. In: Börger E (ed) Specification and Validation Methods, pp 9–37. Oxford University Press
[3] MOST Specification (2005) Media Oriented Systems Transport Specification Rev. 2.4 5/2005. MOST Cooperation, Karlsruhe, Germany
[4] Pavlović D (1999) Semantics of first order parametric specifications. In: Woodcock J, Wing J (eds) Formal Methods '99, LNCS, vol 1708, pp 155–172. Springer
[5] Pavlović D, Pepper P, Smith DR (2004) Colimits for concurrent collectors. In: Dershowitz N (ed) Verification: Theory and Practice: Essays Dedicated to Zohar Manna on the Occasion of His 64th Birthday, LNCS, vol 2772, pp 568–597. Springer
[6] Pavlović D, Smith DR (2001) Composition and refinement of behavioral specifications. Proc 16th Automated Software Engineering Conference, pp 157–165. IEEE Computer Society Press
[7] Pavlović D, Smith DR (2003) Software development by refinement. 10th Anniversary Colloquium of UNU/IIST, Formal Methods at the Crossroads: From Panaea to Foundational Support, pp 267–286. Springer

Embedded Network Processor Based Parallel Intrusion Detection

Hu Yueming

Intel IXA Technology Laboratory
Shanghai Jiao Tong University, China
hu-ym@cs.sjtu.edu.cn

Summary. One of the challenges on Internet intrusion detection (ID) is detecting the intrusion on high-speed networks. To cope with the intrusion on gigabit Ethernet or higher network links, the ID devices must utilize high-speed hardware, parallel structure and efficient algorithms. This paper presents a parallel approach of ID scheme based on network processor, the embedded processor for network devices. In this approach, packets from the network flow through multiple network processors. Within the network processor, multiple network processing engines are used to process the network data in parallel, each network processor/ processing engine detect the packet flow for a subset of intrusion signature. The ID scheme is a MISD parallel mode. Data flow from the network is a single packet flow, and the detection device is a multiprocessor structure with different programs.

Internet intrusion detection (ID) concerns the detection of illegal activities and acquisitions of privileges on the protected subnet. Intrusions can be divided into host based and network based, and so are ID methods. A majority of intrusions from network use a small number of known attacks, which has clear signature[1, 2, 3]. Network based ID is the most concerned because of its "invisible" feature and can be integrated with firewalls to protect the whole subnet.

Network based ID use packet scanning technology to detect intrusion behaviors. One of the challenges on Internet ID is detecting the intrusion on high-speed networks, which need high computing power. Another challenge is to meet the need of detecting the intrusions which may happen in the future, since intrusion and anti-intrusion will be an endless race in the future.

Günter Hommel and Sheng Huanye (eds.), Embedded Systems Modeling – Technology, and Applications, 93–99.

1 Intrusion Detection and Network Processor

Misuse detection utilize intrusion signatures as a way of ID, similar to anti-virus software which extracting virus signature from infected files. Some network intrusion signature can be expressed as a rule, such as source IP address equals destination IP address. Some signatures can be expressed as a series of byte patterns in the packets, which can be detected simply by pattern matching, such as worms and buffer overrun attack. Still, some intrusion can be expressed as a state machine, where the state transition is driven by some intrusion signatures in the packets. Such an ID can be described as graph where vertices represent intrusion state and arcs represent events/conditions required for changing state. The detection of these intrusions consists of a series of steps, each step is expressed as an event detected in packets and is recorded as state transition in a state machine. The event can be detected by pattern matching for a signature. The state transition graph represents the transition of system states along the paths that lead to intrusion state or clear initial state.

The expression of intrusion signature depends on each intrusion behavior. ID system can be implemented as a combination of an expert subsystem, a pattern matching subsystem and a state machine subsystem. Expert subsystem is a rule-based system, which code knowledge about intrusion as a set of *if-then* rules. An expert subsystem for ID can be divided into many parts, each of which has a small subset of knowledge and use this knowledge to scan the packets for detecting intrusion behavior.

ID system needs complex computation and high computing power. Normal ID algorithms need tens or hundreds of memory access, which reduces speed. Packets from the network, when processed, will not be used again. There are few data locality in network packet processing. If general purpose CPU is used for ID, the efficiency will be low. Hardware approaches such as AISC for ID can speedup the process, but have the problem of fixed detection logic, and can not be updated for detecting emerging intrusions in the future. Network processor (NP) is a kind of embedded processor optimized for network applications. NP has high packet processing performance due to a set of programmable packet processing engines (PE) in it. Each PE is a programmable hardware with its own instruction set. The instruction set for the PEs is specifically designed to process packet in the network applications efficiently[4].

NP exploits packet-level parallelism by having multiple PEs. PEs can use RISC structure to be adaptable to different network application, such as micro-engines (MEs) in the Intel's NP product line. Intel's NP product line use one embedded processor XScale core for protocol processing and multiple on-chip MEs for simple packet processing. Each ME is a highly programmable RISC running a single process to avoid context switch. Hardware support is provided for fixed number treads of 1, 4 or 8. Intel's NP has an efficient inter ME communication mechanism, which maximize packet processing throughput. It also has multi-level memory hierarchy which can reduce memory access and hide memory access latencies.

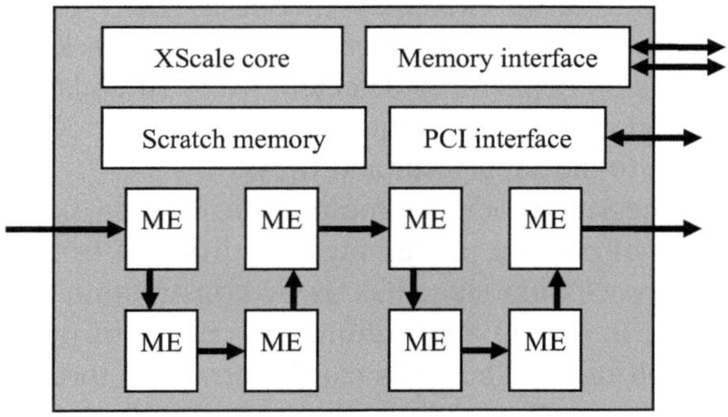

Fig. 1. Main structure of IXP2400 network processor

Intel's IXP2400/2800 NP[5] use XScale core for control plane processing and some complex data plane processing. Multiple MEs are used for wire-speed packet receiving, classification, metering, marking/remarking, queuing, scheduling and packet transmitting, as shown in Figure 1. The IXP2800 has 16 MEs and doubled ME clock rate over IXP2400. Each ME has its own local instruction memory and local data memory as well as a large register file. MEs can share the on-chip scratch memory as well as SRAM and SDRAM space with other MEs and XScale core. MEs can provide performance and programmability for packet processing without OS overhead. It can be used for network based ID. The programmability of network processor makes the of ID system flexible. The ID software can be configured and updated to meet the need of the future.

2 NP Based Parallel ID Scheme

The idea of using NP for intrusion detection is to divide the task into small pieces, with each ME in NP deals with only one subset of detection operation. The XScale core collect information from MEs and doing complex computations such as expert system and data analysis. The database of attack signature is distributed among the MEs. One or two MEs in a NP is used for packet receiving from the network, and extracting the data needed for ID. Each rest ME checks for a fixed number of attack signature with a fixed pattern matching. Packets received from the network flow through

MEs for different signature detection. Different MEs can use different detecting method for signature matching. All the MEs process the packets in parallel. The last ME in the chain simply drop the packet after detection, since ID system is only a "hearing" device in the network.

This system can constantly monitoring the network traffic and detect anomalous traffic pattern, for examples, incoming ICMP broadcast packet and sudden amount of unacknowledged TCP SYN packets to specific servers. When intrusion behavior is detected, ME will update the intrusion packet count or send a message to the XScale core. The XScale core can send alarm message to the supervisor and firewall.

This parallel processing mode is a multiple-instruction-flow single-data-flow (MISD) mode of parallel processing. It's different from pipelining in that each peace of processing stage has its own instruction flow to control the data processing. In the ID application, each peace of processing is independent with each other. The MEs can be arranged in any order, each ME can be inserted or removed from the system without affecting other MEs. The only thing transferred among the PEs is the packet flow or data flow. In normal pipelining, instructions or data flow through the pipeline stages, with only one instruction flow in the system.

Figure 2 depict the structure of this MISD model of parallel intrusion detection, where ID manager is a control plane entity to collect information from ID element and doing data analysis. The ID elements are data plane entities for detecting intrusion signature. This computing model can be mapped easily to IXP2400/2800 NP. The ID manager can be a process running on XScale core of IXP2400/2800 NP. The ID elements are processor elements such as IXP2400/2800's MEs.

Normal computer system don't use MISD mode of parallel execution, because of data dependencies and complex data flow patterns in the application. But in wire-speed network data processing, packets of data come from a network interface in serial, forming a single data flow. Multiple programs running on each PE is independent and easy to write. MISD mode of parallel processing is a natural way of multiple independent processing of data flows from the network.

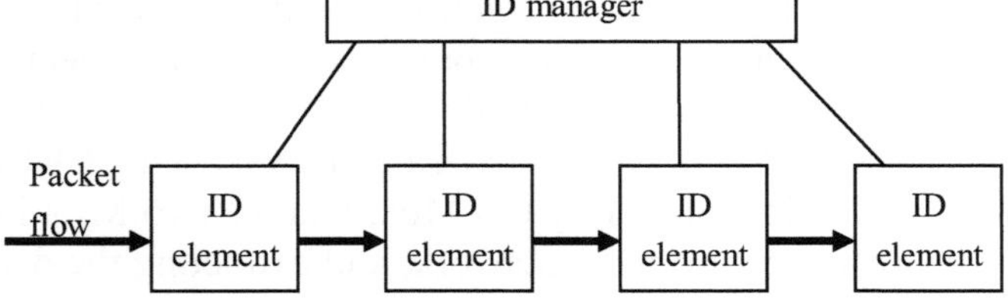

Fig. 2. Structure of MISD intrusion detection

To support the parallel processing scheme, MEs in IXP2400/2800 NP are connected by a fast connection data path and next-neighbor (NN) registers. It can capture every packet in data path and parsing arbitrary sequences of known intrusion patterns. The NN registers in MEs is similar to pipeline register and provides a good way of transfer packet data from one ME to the next. In IXP2400/2800 NP, the MISD mode of execution has the advantage of putting the packet into NN registers and storing intrusion signatures into ME's register or local memory, which greatly reduces memory latencies in normal processing mode. Intel IXP2400 ME has local memory of 3 clock cycle access time and external (off-chip) SRAM memory with access time about 90 clock cycle. The off-chip SDRAM access time of IXP2400 is about 120 clock cycle which is mainly used by XScale core.

Multiple IXP2400/2800 Network processors can be chained with each other to exploit more parallelism of ID application. NPs can be connected in parallel, receiving the same packet flow and detect different intrusions in parallel, as shown in Figure 3.

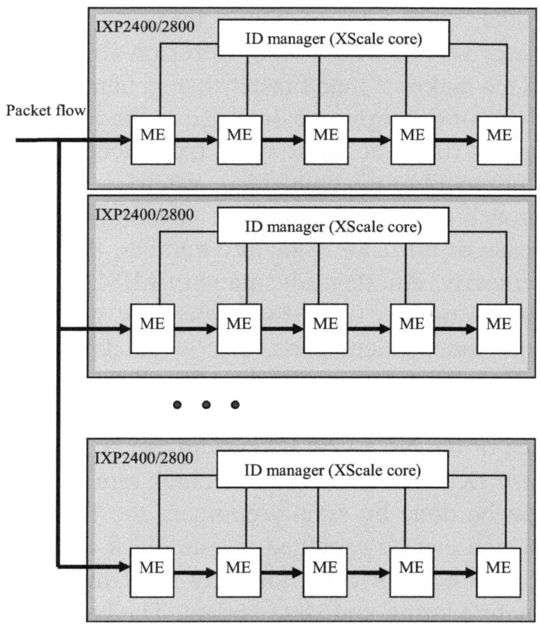

Fig. 3. Parallel connection of NPs for ID

This makes the ID system scalable for more intrusion signatures. The packet output from one NP can also be fed into another NP, for other intrusion investigations. The ID system can be deployed on many positions of Internet and communicate with each other, forming a distributed ID system. In a distributed ID system, the ID managers in each NP should be connected to exchange information through protocols as IDMEF and IDXP.

3 Performance Considerations

In this implementation scheme, each NP need to use one or two MEs for receiving packets from network interface and extract data from each packet. The rest MEs in the NP can be used for ID processing. The extracted data and intrusion signature is stored in registers file or local memory of the ME to avoid external memory access. Each ME in IXP2400/2800 has 256 GPR and 512 data transfer registers. These registers can hold tens to over one hundred intrusion signatures. Some intrusion signatures can be one word of 32 bits, such as IP address. Other intrusion signatures can be a string of multiple words. The whole set of intrusion signature is distributed among the MEs. This makes it much faster than general purpose CPU system which stores part of the signature database in the cache. Another advantage of large register file over cache is its fixed access time to meet the hard real-time requirement of the embedded system.

The MISD approach of ID system has the advantage of linear complexity. With the increase of intrusion signature number, the complexity of ID system increases linearly. Another advantage of MISD mode of execution is reduced complexity of PEs (or MEs), especially when the application can be divided into independent parts, e.g. when doing pattern or table searching where each PE has only to do a small subset of matching or searching.

The performance of the NP based ID system can be estimated as below. Assume that we use IXP2800 with 1.4GHz clock rate of ME. Some intrusion detections can be done by simply compare the IP address, only one ME instruction, which can be executed in one clock cycle, is need to accomplish this. Some intrusion detection may need more instructions because of long signature string and state update. The bandwidth of OC-196 is 9511Mbps, the minimum POS packet length is 49 bytes, the packet arrival rate will be

$$9511\text{Mbps}/49\text{B} = 24.26 \text{ Mpps}$$

So the time budget for processing 1 packets is:

$$(49B/9511Mbps)*1.4GHz = 57.70 \text{ clock cycle}$$

In IXP2800 NP, we can use 2 MEs for packet receiving and 14 MEs for interleaved packet processing. For 14 MEs, the total packet processing time will be 57*14 = 798 cycles. In this time period, pattern matching of about one hundred of intrusion signature is possible. This means that an IXP2800 NP can detect about one hundred intrusion signatures on OC-192 network interface.

For 1Gbps Ethernet, the packet rate for small packets is 1953Kpps, the interval time per small packets is 0.51 micro second, the packet processing time will be 714 cycles, the total packet processing time will be 714*14 = 9996 cycles. We can detect about one thousand intrusion signatures on Gigabit Ethernet with only one IXP2800.

The expansion of NPs can further increase the processing power of the system. Each NP detects fixed subset of attacks in wire-speed. When new intrusion signatures are found in the network, we can update the software of the MEs. When the processing power of existing ME is exhausted, we only have to add new NPs in the system without modifications to the old NPs. This makes it very easy to expand the ID system.

This research work is supported by Intel University Program under grant number of 9078.

References

1. Kumar S, Eugene H. Spafford (1994) A Pattern Matching Model for Misuse Intrusion Detection. Proceedings of the 17th national computer security conference, pp 11-22.
2. Verwoerd T, R. Hunt (2002) Intrusion detection techniques and approaches, computer communications. Vol. 25, pp 1356-1365.
3. T. M. Chen, J. Robert (2004) Worm epidemics in high-speed networks. IEEE Computer, June 2004, pp 48-53.
4. D. E. Comer (2004) Network Systems Design: Using Network Processors. Prentice Hall, Person Education.
5. Intel Inc. (2003) Intel® IXP2400 Network Processor Hardware Reference Manual.

Embedded System Architecture of the Second Generation Autonomous Unmanned Aerial Vehicle MARVIN MARK II

Volker Remuß, Marek Musial, Carsten Deeg, and Günter Hommel

Real-Time Systems and Robotics
Technische Universität Berlin, Germany
{remuss,musial,deeg,hommel}@cs.tu-berlin.de

Summary. This paper presents the second generation of an autonomous helicopter system. Emphasize is put on the architecture of the embedded system and the user-view of the ready-to-fly system.

1 Introduction

MARVIN (multi-purpose aerial robot vehicle with intelligent navigation) Mark II (M2) is an autonomous unmanned aerial vehicle (UAV) that was build in 2005 (see Fig. 1).

Its predecessor [1] was developed from 1997 to successfully win the International Aerial Robotic Competition (IARC) [2] in 2000. From 2002 to 2005 it was used in cooperation with other UAVs in the European research project COMETS [3].

Fig. 1. MARVIN mark II in autonomous flight

Günter Hommel and Sheng Huanye (eds.), Embedded Systems Modeling – Technology, and Applications, 101–110.

2 The M2 Autonomous Helicopter System

The M2 airframe is a commercially available helicopter made in Germany by Aero-Tec [4]. Its main feature is the sealed, full metal, single-stage main rotor gear which is mounted directly on the engine, building a self-supporting unit (see Fig. 2).

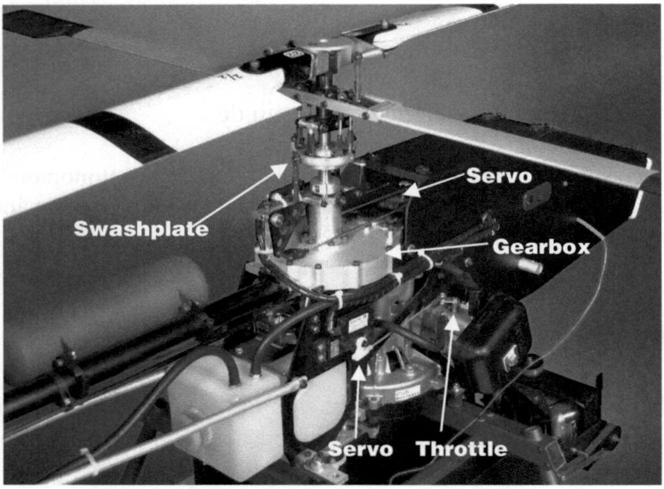

Fig. 2. Airframe's mechanics

Its additional features make it a good choice for an autonomous helicopter airframe. The setup acquired by TUB is equipped with a magnetic sensor on the main rotor axis to be able to measure the rotor's rotation per minute (rpm), a high-stand landing gear with mounting capabilities and an electric starter. The remote control (RC) equipment and gyro uses standard parts produced by Graupner [5]. The engine is a Zenoah G230RC two-stroke 23 cm^3 gasoline-oil-mix engine with a maximum power output of 1.84 kW at 12500 rpm [6].

The helicopter has a two-bladed main rotor with a diameter of 1.84 m and a two-bladed tail rotor with a diameter of 0.34 m. The tail rotor is coupled to the main rotor via a fixed-ratio gear. The main rotor's rotary speed is actively fixed during flights at about 1350 rpm. Changes in lift, which are needed to move the helicopter, are produced by changing the main rotor blades' pitch. This is done symmetrically (collective pitch) to change the lift and asymmetrically (cyclic pitch) to accelerate in a direction. To set the pitch relative to the helicopter a so-called swashplate (SP) is used. It is moved by three servos. The tail rotor is used to compensate for the main rotor and engine torque and change the heading of the helicopter. Its (collective) pitch is controlled by a single servo. A fifth servo moves the throttle. The SP and the throttle's servo can be seen in Fig. 2. The three SP servos are inside the dark front box of the helicopter, but their linkage is visible.

2.1 Remote-Controlling a Model Helicopter

The servos are controlled by pulse width modulated (PWM) signals where the PW is proportional to the servo anchor's position. In radio controlled flight the servos are driven by the RC receiver as received from the RC. The RC pilot does not control directly the servos but uses a higher-level approach. The pilot has two 2-way sticks, one is controlling the cyclic pitch, which has 2 degrees of freedom (DOF) and the other one is controlling the collective and the tail rotor pitch with 1 DOF each. The throttle is usually fixed to a certain base level. The RC derives the individual servo signals from the pilot's commands. The RC also compensates for some couplings, i.e. it raises throttle when the collective pitch is raised. In case of the M2 airframe the level of abstraction is very low because there is a dedicated collective pitch servo that moves the other two (cyclic pitch) servos.

Mostly, a RC pilot uses a 1 DOF gyroscope to control the tail rotor pitch to prevent unwanted rotation around the main rotor axis. All of these automatisms have to be adjusted for each particular set of helicopter model, servos, rotor blades, fuel, weather and so on. The RC transmits unidirectional, so there is nothing but visual feedback from the flying helicopter.

2.2 From a Remote-Piloted Helicopter to an Autonomous UAV

To be able to fly a model helicopter autonomously one has to control the helicopter automatically. The automatic control can be engaged on any point between in-front-of-the-remote-control to just-before-the-servos. Due to automatisms mentioned above, the decision was taken to go to the lowest level and directly interface the servos on the helicopter and to spare out any RC automatisms. A microcontroller (MC) was used for this purpose. The decision to switch between autonomous- and remote-controlled modes is taken by the MC's program but can be overridden by a lever on the RC for security reasons. Switching back to manual flight is always the last resort in case of system problems.

Since two different batteries for the microcontroller and the RC receiver and servos are used, the hardware falls back to remote-controlled flight if the MC is running out of battery. In addition the MC is able to monitor the signals that are sent to the servos during a remote-piloted flight thus they can be logged. In earlier versions of MARVIN all the servo and receiver interfacing was performed using dedicated custom-built hardware and the MC directly. In M2, this is no longer necessary since a receiver system is available that comes with this functionality built-in (see below). M2 can also start its engine autonomously using the electric onboard starter.

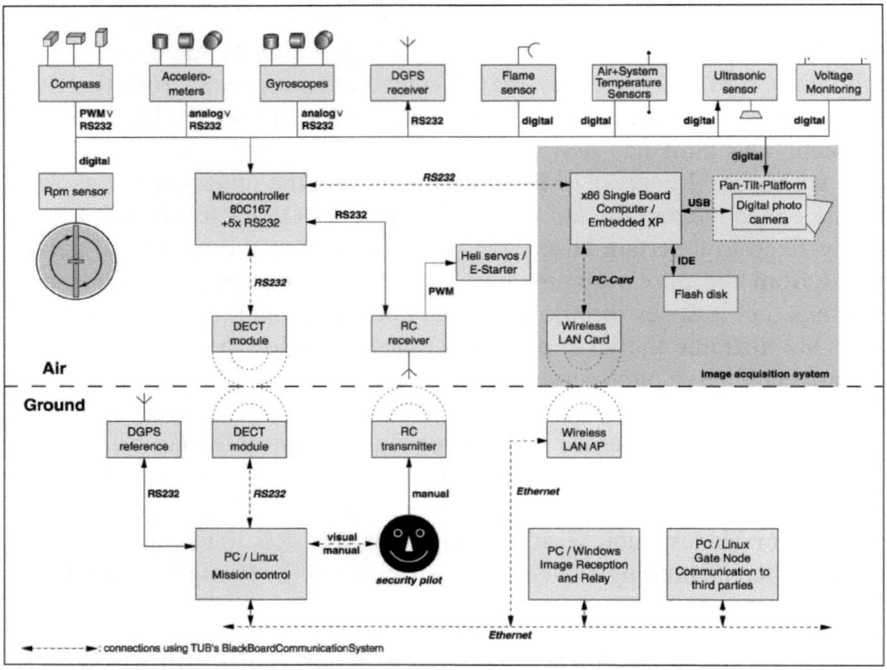

Fig. 3. Complete autonomous M2 system incl. ground base and optional image acquisition system

The Infineon C167 MC is the only CPU on board the helicopter used for autonomy. Therefore, all the needed sensors are connected to it and it executes all programs for control purposes, sensor data acquisition and calculations. To fly the helicopter autonomously the control program in the microcontroller needs information about the current position and the attitude of the helicopter. Additionally, the current main rotor rotations per minute (rpm) are needed to control the engine's throttle and a sensor for the distance to the ground for landing purposes.

The system uses a Differential Global Positioning System (DGPS) to get accurate and frequent position information and a combination of three rotational speed, three acceleration and three magnetic field sensors for movement detection. A fusion algorithm is used to calculate the helicopter's attitude from raw sensor data [7]. The main rotor rpm is measured magnetically at the rotor axle and the distance to the ground by an ultrasonic sensor. Each of these sensors is directly connected to the MC. To add the D to GPS a stream of correction data from a base station is needed. In case of M2 it is embedded in the normal data transfer between the base station and the helicopter and then transferred through the MC to the on-board GPS using TUB's real-time communication system (CS) [8, 9].

Figure 3. depicts the full M2 embedded architecture including a view of ground station which consists of several networked computers using the same CS in all nodes. The architecture includes the image acquisition system as well as all components that are needed for autonomous flight.

2.3 Hardware Components Realisation

The decision was taken to go for a flexible and comfortable approach (for research, development and maintenance) and to attach a box beneath the helicopter that can contain a variable number of computer systems, sensors, power supplies, and batteries. Another benefit of this design is the quick replaceability of the airframe for repair issues. The box is completely cut from carbon fiber sheets besides the aluminium back that is used for most connectors and as power supply's heat sink. Figure 4. depicts M2's integrated electronic box including image acquisition system. Most components of the M2 system are visible.

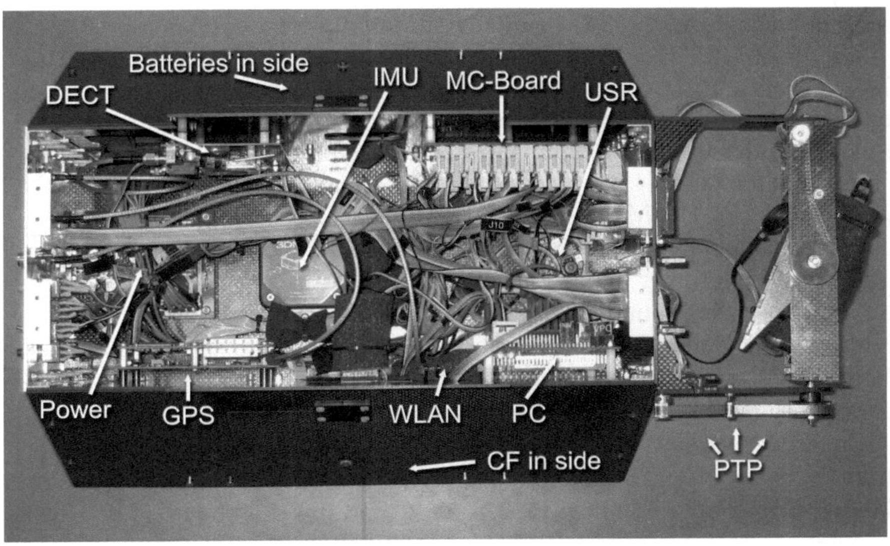

Fig. 4. M2's integrated electronics

The M2 uses a DECT (Digital Cordless European/Enhanced Telephone) OEM module by Hoeft+Wessel [10] as a transparent serial link for communication with the base station. To ease the integration an end-user product was used (HW8615). The good thing about this DECT modem is that it comes including an external antenna connector and an own power supply that fulfils the specification of the OEM module. Compared to the formerly used Siemens M101 Data these modules were easier to integrate and use, but their actual performance was less stable.

The box in Fig. 4 shows an inertial measurement unit (IMU) GX1 from Microstrain [11]. This IMU integrates all needed sensors (rotation, acceleration and magnetic in 3D) and interfaces the MC via RS-232. While the integration of all these sensors in a single device seems a very nice idea at a first glance it turned out that for M2 it was not possible to find a mounting place that was suitable for all the sensors at the same time. The accelerometers and gyroscopes are best placed decoupled below the helicopter's centre of gravity while the magnetic sensors are heavily disturbed by the two-stroke engine's induction coil and are best placed as far away as possible. In M2's case "far away" is the helicopter's tail section. So it was necessary to add a compass to the Microstrain when it was mounted as in the picture above or to use the Microstrain as compass only (mounted on the tail) and use another IMU in the box.

Heart of the system is the custom made MC-motherboard that contains an integrated Infineon C167 MC module by Frenzel+Berg [12], temperature and voltage monitoring and a lot of lockable connectors. The C167 is a 16-Bit microcontroller of good connectivity including Capture/Compare-Units, AD-Converters and GPIO. The Frenzel+Berg module comes with external Flash, external RAM and 4 additional RS-232 serial ports.

The used GPS is a high accuracy OEM-4G2L by Novatel [13]. This GPS is used in differential mode accompanied by a second, older Novatel receiver (OEM-3) on ground. A second existing M2 helicopter is also using the older OEM-3 model. Both systems show a similar accuracy of 1 cm while the OEM-4 has a maximal position update rate of 20 Hz and the OEM-3 of 5 Hz. The M2 system uses 5 Hz or 10 Hz, respectively.

Directly integrated into the base plate of the box facing downwards is an ultrasonic rangefinder (USR) produced by SmartSens [14]. It is mainly used to detect the ground during autonomous landing. Four high-bright red light emitting diodes (LEDs) in the base plate are used to give feedback about the current flight mode (i.e. helicopter turning, hovering, start accelerating, or landing) when the helicopter is autonomous.

The power supply unit at the aluminium sheet takes a variable voltage from the batteries and converts it to 5 V and 3.3 V using mostly switching DC-DC-Converter modules PL10S produced by Lambda [15]. These modules produce a well regulated, comparatively low-noise output suitable for all digital systems. In addition, there are very low-noise linear regulator for analogue systems especially sensors and AD-converters. The complete M2 system is powered using lithium-polymer rechargeables from Kokam [16]. A 11.1 V, 3 Ah, 235 g battery pack powers the system including the image acquisition system for at least 90 minutes. The M2 can carry up to 3 batteries which are automatically balanced during discharge.

Not inside the box is the DDS-10 RC-receiver produced by ACT [17]. Since the receiver is used by the helicopter for manual RC it is installed in the helicopter's nose and powered by a dedicated battery. This receiver has a unique feature called diversity synchro-link system (DSL-S) which enables the C167 to control the servos via RS-232 or read the values as sent by the RC. A master-slave setup still enables a security override switch at the RC. The receiver's design is quite robust, it falls automatically back to RC when the microcontroller stops sending datagrams but it also keeps on using the microcontroller's data in case the RC is failing or out of reach. Additionally, the receiver monitors its current reception strength and battery voltage.

Also in the box is the image acquisition system's embedded PC. This PC is a PC-104+ compliant, embedded National Semiconductor (now AMD) Geode system-on-chip SB-i686 by Compulab [18]. It is operated with embedded Windows XP to drive the digital still camera that is connected via USB and boots from a compact flash card (CF) attached via IDE. Its pc-card slot carries a wireless LAN card for image transfer to the base station, but the wireless LAN can also transmit all data transferred from the MC to the base station in case of a DECT failure. The embedded PC is not necessary for autonomous flight. Together with the WLAN, CF, the pan-tilt-platform (PTP) and the camera it is just payload.

Another mission-specific sensor is the flame sensor hidden in a small external white box at the front most part of the helicopter. This fire or more exact a flame sensor is a Hamamatsu UV-Tron R2868 [19]. It is very sensitive in detecting flames but is absolutely insensitive to other sources of heat and light, e.g. the sun. The sensor is detecting the intensity. In the M2 system the sensor is masked to a narrow opening angle in the horizontal plane and it is looking to the front, in flying direction. In this configuration it can be used to avoid flying into fires and to triangulate them.

2.4 Base Station

The M2 base station consists of several components but the salient program is the Mission Control, mainly because it has a GUI (see Fig. 5.) and represents the base station's user-view.

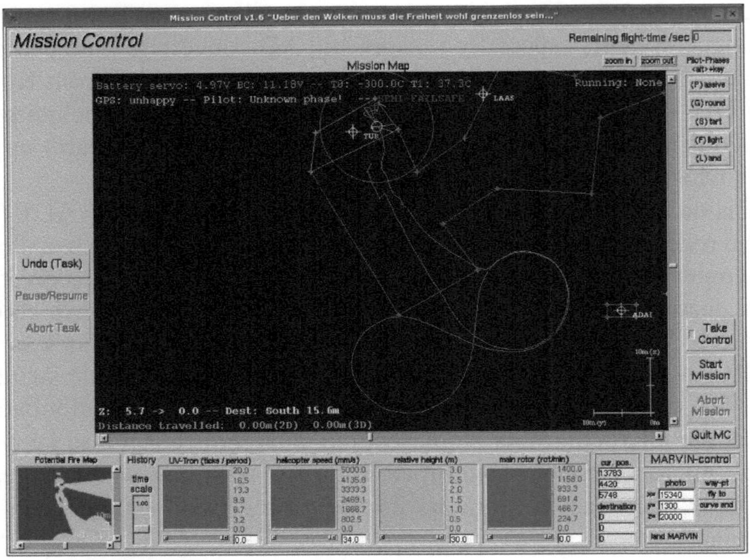

Fig. 5. Mission Control Program (view after a real flight)

This program can be used to plan and survey a mission flight including definition of waypoints to fly to and giving the command for start and abortion if necessary. It can display some basic digital map data, the helicopter's position and heading as well as the flown path. Some status information and flight parameters are shown as well. Behind the scenes the MC sends atomic commands to the helicopter systems that are then executed independently. Terminating the MC does not end the autonomous flight.

2.5 MARVIN's Flight

Flying a path for MARVIN is always flying from one three-dimensional waypoint *(i–1)* to another *(i)*. MARVIN has two different flight modes. One is *straight line* and the other *curved.*

When flying *straight line* mode, the system takes the shortest path from waypoint *i–1* to *i* usually nose ahead. If *i* is reached, it stops, turns toward *i* + *1* and accelerates. MARVIN is flying the way with a given (maximal) cruise speed.

When flying *curved* mode, the helicopter keeps the cruise speed until it reaches the last way-point and does not stop at every intermediate waypoint. To achieve that, the control switches from waypoint *i* to *i* + *1* as soon as the helicopter is so close to *i* that it would need to slow down to stop at point *i*. The result is a smooth transition between waypoint. Figure 6. depicts a simulated flight that shows both modes in comparison.

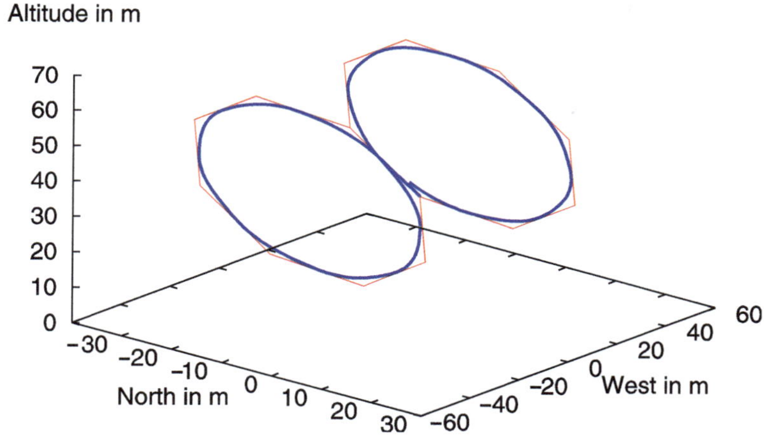

Fig. 6. Simulated flight

It is possible but rarely used to fix the direction of MARVIN's nose while flying. The mission flight speed is typically 2 to 5 m/s, but successful flights have been performed with a velocity of 14 m/s.

The flight range of the helicopter in its current state is about 300 m due to two issues. Firstly, as long as a human pilot should be able to recover the helicopter in case of an emergency it has to be somewhere in the middle of the area because the visual range is about 150 m. Secondly, DECT and WLAN are officially limited to 300 m but this is solely a monetary issue which could be solved easily.

The maximum flying height is an even more difficult question. A maximal ascending speed of 2 m/s can be reached, but flying beyond a height of 50 m is risky since this is the maximum height that a human pilot is able to fly. Disregarding the safety factor, the reachable height is 300 m as well due to the radio components.

Safety is an issue. It may seem a pretty small helicopter but if one looks at the technical data and think about it one will see, that there are rotating blades with 1350 rpm, having a diameter of 1.84 m and weight 0.25 kg each. Therefore, the tip of the blade has a speed of more than 450 km/h and each blade is pulling with a force of about 2800 N outwards.

3 Conclusion

In this paper the autonomous helicopter's embedded architecture has been presented in detail including some very specific problems and solutions. Additionally, the user-view has been presented with the goal to give an integrated picture of the real life performance of the system.

Acknowledgements

Part of this work has been funded by the European Commission in the Information Society Technologies Program under contract COMETS IST-2001-34304.

References

[1] M. Musial, U. W. Brandenburg, G. Hommel. *MARVIN – Technische Universität Berlin's flying robot for the IARC Millennial Event.* In Proc. Symposium of the Association for Unmanned Vehicle Systems 2000, Orlando, Florida, USA, 2000.
[2] Association for Unmanned Vehicles International. *International Aerial Robotics Competition: The Robotics Competition of the Millennium.* http://avdil.gtri.gatech.edu/AUVS/
[3] *COMETS, Real-time coordination and control of multiple heterogeneous unmanned aerial vehicles,* IST 34304, 5th Framework Program, http://www.comets-uavs.org, EU, 2002
[4] Aero-Tec, CB-5000, Germany, http://www.aero-tec-helicopter.de/
[5] Graupner GmbH & Co. KG, Germany, http://www.graupner.de/
[6] KOMATSU ZENOAH CO., Japan, http://www.zenoah.net/
[7] M. Musial, C. Deeg, V. Remuß, G. Hommel: *Orientation Sensing for Helicopter UAVs under Strict Resource Constraints.* Proc. First European Micro Air Vehicle Conference EMAV 2004, Braunschweig (Germany), July 13-14, 2004.
[8] V. Remuß, M. Musial, *Communication System for Cooperative Mobile Robots using Ad-Hoc Networks,* Proceedings of the 5th IFAC Symposium on Intelligent Autonomous Vehicles, 2004, Lisbon, Portugal, Elsevier Science, ISBN 008-044237-4
[9] V. Remuß, M. Musial, and U. W. Brandenburg. *BBCS – Robust communication system for distributed systems.* Proc. IEEE Intern. Workshop on Safety, Security and Rescue Robotics (SSRR 2004), May 2004. ISBN 3-8167-6556-4
[10] Höft & Wessel AG, HW8615, www.hoeft-wessel.com
[11] Microstrain, GX1, USA, www.microstrain.com
[12] Frenzel + Berg Elektronik, Germany, www.frenzel-berg.de
[13] Novatel Inc., Canada, www.novatel.ca
[14] SensComp Inc., USA, www.senscomp.com/600smartsensor.htm
[15] Lambda, Europe, www.lambdaeurope.com
[16] Kokam, Korea, www.kokam.co.kr/english/index.html
[17] ACTeurope, Germany, www.acteurope.de
[18] CompuLab Ltd., Israel, www.compulab.co.il
[19] Hamamatsu Photonics, Japan, www.hamamatsu.com

Middleware for Distributed Embedded Real-Time Systems

Marek Musial, Volker Remuß, and Günter Hommel *

Real-Time Systems and Robotics
Technische Universität Berlin, Germany
{musial,remuss,hommel}@cs.tu-berlin.de

Summary. This paper identifies the middleware requirements typically connected with heterogeneous distributed embedded real-time systems and presents the BBCS middleware developed by the authors to meet these requirements, with special focus on bandwidth scheduling and the user's view.

1 Introduction

In the design of any distributed embedded system, the task of communication between different sources and sinks of information has to be solved. This always involves both dedicated hardware and dedicated software to provide a suitable communications service. For the corresponding software layer, the term *middleware* (see e.g. [1]) has been coined, indicating that communications occupy an intermediate role between the local operating system of every system module and the distributed "application" governing the distributed system.

Whenever the focus is on embedded distributed real-time systems, it turns out that a relatively constant set of requirements arise with respect to suitable middleware. Notably, both the real-time system properties and the hardware diversification met in such systems tend to be poorly addressed by available middleware products. Of course, attempts are being made to address these issues "once and for all", e.g. through the *Common Object Request Broker Architecture* (CORBA) [4] with its real-time extension [3], but these suffer heavily from their out-of-hand complexity, and they still do not specifically address the real-time requirements of the low-level data exchange procedures.

The typical set of requirements mentioned above includes at least the following list of aspects:

1. There are very different pieces of information to be transmitted between pairs of system modules. The difference pertains to the size of information units, their

* This work has been partially supported by the COMETS project (European Commission, IST 2001-34304, [2]).

Günter Hommel and Sheng Huanye (eds.), Embedded Systems Modeling – Technology, and Applications, 111–120.

production (or update) rate, and the upper bound on the delay of their being relayed, as imposed by the real-time guarantees to be met by the system.
2. The available outgoing communication bandwidth from every module needs to be shared between these different pieces of information transmitted by the module. This sharing, in turn, affects the resulting transmission delays.
3. A great fraction of the information to be exchanged consists of *state variables*. State variables shall denote data portions of a fixed size, with only the most recent content of the variable being of relevance. I.e., in the exchange of state variable contents, the loss of data (or failure to schedule a transmission within a certain time window) is fully cured by the next successful transmission. Therefore,any use of flow control or means of retransmission would be counterproductive.
4. The middleware should be usable with a great variety of processing hardware, operating systems, and communication media, in order to enable the design of heterogeneous distributed real-time systems in a most transparent way. As this includes the implementation on microcontrollers, a relatively light-weight middleware system is obligatory.

The rest of this paper introduces and describes the *Blackboard Communication System*, BBCS, a middleware system meeting a superset of above requirements. Certain aspects of BBCS have already been described elsewhere, e.g. [8, 7]. This paper specifically addresses the bandwidth scheduling algorithm used by BBCS formally and depicts the usage of BBCS for state variable transmission.

The subsequent section summarizes all features provided by BBCS. Section 3 covers BBCS's algorithm for bandwidth scheduling and the resulting real-time guarantees. Section 3 explains the application of BBCS to state variable transmission by means of examples.

2 Features

This section gives a brief overview of the design features implemented in the BBCS middleware. Technical details are mostly left out here.

2.1 Architecture

A distributed system utilizing BBCS consists of *nodes* and point-to-point communication *channels*. This does not impose any unsuitable restriction, because

1. broadcast-type communication media can still be used to establish multiple point-to-point channels, and
2. broadcast-type communication modeling would have been fairly undesirable due to the resulting medium-wide scheduling constraints.

The underlying *blackboard* architecture denominates some virtual memory storage that is transparently synchronized between the BBCS nodes via the BBCS protocol (see figure 1). With respect to state-variable type data, this provides almost complete abstraction from distributedness.

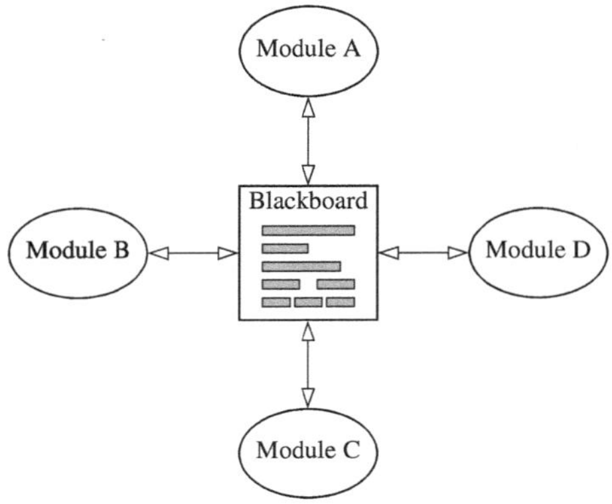

Fig. 1. Blackboard architecture view of BBCS.

2.2 BBCS Protocol

Every BBCS channel fulfills the requirements listed in section 1. In excess of supporting state-variable class communication, BBCS also offers a service to transmit reliable data streams with flow-control within the channel's data traffic. The distinguishable data portions are called *slots*. Each slot thus constitutes either a certain state variable or a data stream and can be assigned a fixed guaranteed fraction of a channel's bandwidth. The BBCS protocol subdivides the exchange of data into fixed-size packets, called *stripes*. The size of these stripes, a configuration option, defines the minimum granularity of bandwidth scheduling. State variables may occupy much more memory than the stripes' size. Then, BBCS will use as many stripes as required to transmit a new content of the state variable, but the update will be presented as atomic to the user, in order to preserve the non-distributed abstraction. For every slot's contents, there can be at most one node in the network acting as the *data source*, but as many nodes as desired may read the slot's data (as *data sinks*).

2.3 Shortest-Path Routing

In a BBCS network with many nodes and channels, there might be more than one possible transmission path between a data source node and a corresponding sink node. BBCS performs dynamic shortest-path routing throughout the network. Dynamic here means that an alternative route is set up automatically if one of the channels suffers a temporary failure and an alternative route of working channels still exists. The initial shorter route will be re-enabled as soon as the faulty channel resumes its operation.

2.4 Platform Independence

In order to achieve maximum independence of operating systems and communication devices, the BBCS's design has been split into a big platform- and device-independent main part and a very small platform-dependent part. The latter is called *platform abstraction layer* (PAL), providing only a very small set of services (basically transmission and reception of raw data packets, memory management, and real-time clock). So far, PALs have been successfully implemented for and BBCS utilized on the following set of platforms: Solaris, Linux, Windows, Windows CE, generic POSIX, QNX, Infineon SAB80C166/167 microcontroller family, Atmel ATmega128 microcontroller, and LabWindows CVI real-time controllers. Low-level communication devices and protocols supported by the PAL implementations so far are: TCP, UDP, and RS232 serial links.

2.5 Practical Use

BBCS was developed and served as the primary middleware in the European Commission IST project COMETS [2] and for TU Berlin's UAV MARVIN [5, 6].

3 Bandwidth Scheduling

This section presents the bandwidth scheduling scheme and algorithm used by BBCS.

As explained in section 2, bandwidth scheduling is performed for each sending node and channel with *stripe* granularity. In order to guarantee deterministic bounds, a *static* transmission schedule is generated whenever the set of bandwidth requests for the node and channel in question changes. This static schedule consists of a list of 2^K slot numbers $\langle s_i \rangle$, $0 \le i \le 2^K - 1$, defining the order of slots to be allowed to transmit exactly one stripe packet. Whenever the end of the schedule is reached, it is applied again from its beginning, and so on. Thus, the effective resolution of the schedule, in terms of the minimum bandwidth fraction b_{min} to be assigned to a slot, is determined by the parameter K as:

$$b_{\text{min}} = 2^{-K} \tag{1}$$

The next section describes the basic algorithm used to process the bandwidth requests, called *static binary tree bandwidth scheduling* (SBTBS), and proves its correctness. The subsequent section 3.2 depicts the obtaining of the $\langle s_i \rangle$ schedule from SBTBS's output. Section 3.3 then proposes and proves the resulting guarantees on transmission delay.

3.1 Scheduling Algorithm

The SBTBS algorithm constructs a binary tree T, the leaf nodes of which are marked with a slot number or the special tag *EMPTY*. Initially, T consists only of its root node, marked *EMPTY*. Every node represents $1/2$ of the bandwidth portion represented by its parent, while the root node represents 100% of the bandwidth available.

SBTBS further utilizes a list $\langle f_i \rangle$, $0 \le i \le K$, of pointers to nodes of T. f_i points to the only empty node at level i of T, or is *NIL* if there is no such empty node. Initially:

$$f_0 := \text{root}(T) \tag{2}$$
$$f_j := NIL \qquad (1 \le j \le K) \tag{3}$$

Below, SBTBS is presented for the reservation of a bandwidth fraction of 2^{-i} for slot s, with $0 \le i \le K$. 2^{-K} is the resolution of the schedule and constitutes the smallest fraction reservable. Fractions that are no powers of two can obviously be reserved by simply using the same procedure repeatedly for every 1-bit within the fraction's binary representation.

1. Determine the smallest empty node n in T big enough for the requested fraction:

$$k := \max\{j | i \ge j \ge 0 \wedge f_j \ne NIL\}$$
$$n := f_k \tag{4}$$

   If no such k can be found, more than 100% of the total bandwidth is being requested. In this case, return with failure.
2. Forget the corresponding empty-node pointer:

$$f_k := NIL \tag{5}$$

3. While $k < i$ (i.e. while node n represents more relative bandwidth than requested), split its bandwidth fraction among two new empty child nodes, recording the "right-hand" child in the list $\langle f_i \rangle$, and proceed with the "left-hand" child:

$$k := k+1$$
$$m := \text{add\_child}(n, EMPTY)$$
$$f_k := \text{add\_child}(n, EMPTY)$$
$$n := m \tag{6}$$

4. Now, n is an empty node with the correct size. Mark n with slot number s. Return with success.

The correctness of SBTBS can be proven by showing two propositions:

1. Between any two requests, the nodes listed via $\langle f_i \rangle$ together exactly represent the total bandwidth fraction not yet reserved.
2. SBTBS completes successfully iff a sequence of permissible bandwidth requests does not exceed 100%.

Proof of prop. 1: Proof by induction over the number of requests. Initially, the root node is tagged *EMPTY*, pointed to by f_0, and represents 100% of the available bandwidth. This anchors the induction. Induction step over one execution of SBTBS: Node f_k was initially empty because of step 1 and is removed from $\langle f_i \rangle$ in step 2. For each execution of the *while* body in step 3, the remaining portion such removed from $\langle f_i \rangle$ and not yet assigned is split in two, one half newly included into $\langle f_i \rangle$ and the other half kept recorded in n, also a portion unassigned but not currently present in $\langle f_i \rangle$. Here, it is important to note that all pointers f_j, $k+1 \le j \le i$, with k from

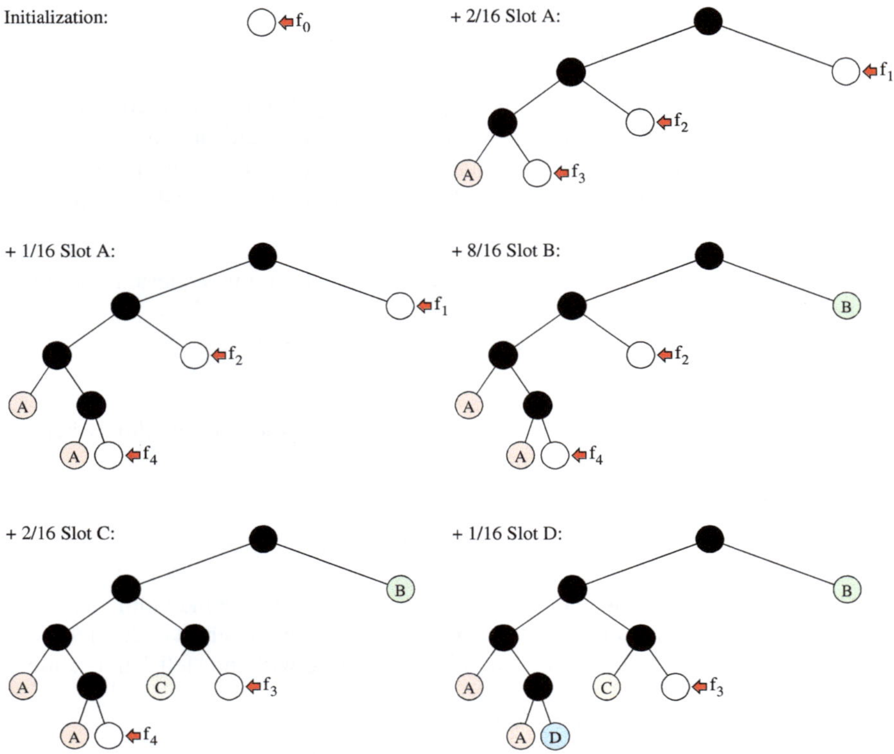

Fig. 2. Example run of the SBTBS algorithm, $K = 4$, $b_{\text{min}} = 1/16$.

step 1, were *NIL* according to the calculation of k. Finally, the node referred to by n is assigned to the requesting slot, thus correctly no longer listed in $\langle f_i \rangle$. □

Proof of prop. 2: Whenever a "suitably big" free node is found in step 1, SBTBS succeeds. This is correct because the node found in this way was unused, by prop. 1. Otherwise, by prop. 1 and step 1, all single remaining free nodes are smaller than the request 2^{-i}, which means that their sum is also smaller than 2^{-i}, because:

$$\sum_{j=i+1}^{K} 2^{-j} < \sum_{j=i+1}^{\infty} 2^{-j} = 2^{-i} \qquad (7)$$

Therefore, if step 1 returns failure, the request must have exceeded the total remaining bandwidth portion. □

Figure 2 visualizes the execution of SBTBS for the following request list, from left to right then top to bottom:

Slot	A	B	C	D
Bandwidth request	3/16	8/16	2/16	1/16

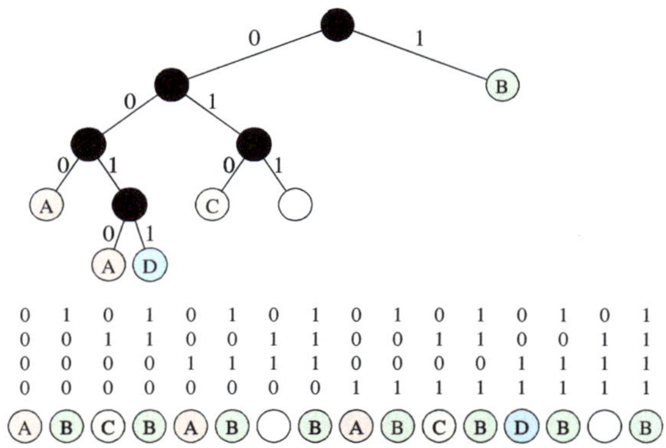

Fig. 3. Derivation of schedule s_i from the SBTBS output.

Please note that slot A is processed through two subsequent requests, representing $3/16 = 2 \cdot b_{\min} + 1 \cdot b_{\min}$. The arrows depict the list $\langle f_i \rangle$ in every state. Finally, one empty node representing the unassigned fraction of $2/16$ is left.

3.2 Derivation of Schedule

It remains to be seen how the schedule in the $\langle s_i \rangle$ representation is obtained from the final tree T. It is essential to spread every bandwidth assignment over the schedule as evenly as possible, while the nodes of T show the most compact distribution – which is in fact just the opposite extreme. The "most even" distribution is easily obtained by:

1. Labeling the two outgoing edges of every node with 0 and 1, respectively.
2. Setting s_i according to the tag of the leaf node that is reached from the root, by following the edges as determined by the binary representation of i, from least to most significant bit (LSB to MSB).

Figure 3 depicts this for the tree from figure 2. The "empty" entries s_6, s_{14} will be skipped on executing the schedule, in favor of the effective bandwidth provided to each of the slots.

3.3 Real-Time Bounds

Proposition 3: Let $d_{\max}$ denote the maximum delay possible, in stripe transmissions, between two subsequent transmission times of a stripe of slot n with an assigned bandwidth fraction r, according to the schedule generated above. Then:

$$d_{\max} = 2^{-\lfloor \mathrm{ld}\, r \rfloor} \tag{8}$$

Proof of prop. 3: For a total relative bandwidth r, the biggest fraction of r assigned through a single call of SBTBS is 2^b, with $b = \lfloor \mathrm{ld}\, r \rfloor$ denoting the exponent of the most significant bit of the binary representation of r. Therefore, the path to the corresponding node of T is $-b$ edges long, and has thus been periodically selected every 2^{-b} positions during the construction of $\langle s_i \rangle$. This already implies (8). □

Please note that this $d_{\max}$ is equal to the theoretical optimum of $1/r$ for proper powers of two, $r = 2^{-b}$, and always better than the theoretical optimum doubled:

$$1/r \leq d_{\max} < 2/r \tag{9}$$

4 Usage

This sections provides an overview of using BBCS in an application by means of examples. These examples deal with state-variable type communication, reliable data streams are outside the scope of this paper (their usage does not significantly differ from usual `read` and `write` operations).

The following two sections deal with different example scenarios, transparent periodic access on one hand and occasional data push on the other. All examples are basically written in C. Portions regarding the initialization of BBCS and the communication channels will not be considered.

4.1 Transparent Periodic Access

Transparent periodic access means that the application can use a state variable just as if it were fully local. A convenient example for this kind of situation would be a simple controller that gets the current speed vector $v \in \mathbb{R}^3$ of some unmanned aerial vehicle (UAV) as its input and adjusts the UAV's elevators in order to keep the vertical speed (say, v_2) close to zero. The input and output variables `vel` and `elv` shall be defined, in fixed-point notation, in this way:

```
long vel[3];
long elv;
```

Here, it can be assumed that `vel` is measured and `elv` applied by another system node than the controller is running on.

Now, the usage of BBCS within the controller node mainly consists of:

1. Some initialization code, binding a BBCS slot to the two "blackboard" variables:

   ```
   bb_add_slot(VEL_SLOT, 0, sizeof(vel), &vel, 90);
   bb_add_slot(ELV_SLOT, 0, sizeof(elv), &elv, 30);
   bb_auto_get(VEL_SLOT);
   bb_auto_put(ELV_SLOT);
   ```

   Here, `VEL_SLOT` and `ELV_SLOT` are just constants to identify the slots. 0 is the flag defining the two slots declared here to use state-variable semantics (as opposed to reliable data-stream mode). The size and address expressions bind the user variables to the corresponding slots, and 90 and 30 are the bandwidth fractions (in 1/1000) to be assigned to the slots. Note that the `vel` variable occupies

more memory than `elv`, so it might be reasonable to assign bandwidth correspondingly.
`bb_auto_get` and `bb_auto_put` instruct BBCS to transparently synchronize the slots' user variables whenever an update is received, or whenever bandwidth is available, respectively.

2. The control loop, where there is no trace of distributedness:

   ```
   elv = -42.17 * v[2];
   ```

   (This is a simple P controller with some arbitrary gain, but this detail does not really matter here.)
3. Somewhere in the main loop of the controller process, instruction to BBCS to perform all pending synchronization:

   ```
   bb_sync();
   ```

   It has been a deliberate design goal that BBCS operates synchronously to the user application, requiring `bb_sync` for the explicit invocation of synchronization activities. This simplifies user applications considerably.

4.2 Occasional Data Push

Occasional data push means that updates of a slot variable are relatively seldom, and the update events need to be handled explicitly. A possible example would be the changing of a camera's focal length, particularly if the corresponding zooming operation requires significant extra time. Using a variable

```
int f;
```

for the focal length in mm, the code might look like:

1. For the initialization, assigning 1% of the available bandwidth:

   ```
   bb_add_slot(F_SLOT, 0, sizeof(f), &f, 10);
   ```

2. At the *sending node*, a human user might trigger the change of focal length, resulting in a function call:

   ```
   void changeFocalLength(int value)
   {
     f = value;
     bb_put(F_SLOT);
   }
   ```

   At the *receiving node*, the resulting slot updates can be detected in the camera control loop like this:

   ```
   if (bb_get(F_SLOT) > 0) {
     setCameraFocalLength(f);
   }
   ```

3. Also in this case, both nodes must periodically perform:

   ```
   bb_sync();
   ```

5 Conclusion

In this paper, the BBCS light-weight middleware for heterogeneous distributed embedded real-time systems has been introduced. It has been pointed out that BBCS has been designed according to *typical* communication requirements arising in this kind of systems. The core of BBCS with respect to its real-time behavior is determined by the SBTBS algorithm for the calculation of static bandwidth schedules. This algorithm has been explained in detail and shown to be correct and close to optimal with respect to its real-time guarantees. The application of BBCS to the propagation of state variables has been illustrated by means of examples.

References

1. Middleware Resource Center. http://www.middleware.org.
2. COMETS consortium. Real-time coordination and control of multiple heterogeneous unmanned aerial vehicles, 2002–2005. IST-2001-34304, http://www.comets-uavs.org/.
3. Object Management Group. Real-time CORBA, version 2.0. http://www.omg.org/, 2003.
4. Object Management Group. Common object request broker architecture (CORBA/IIOP), version 3.0.3. http://www.omg.org/, 2004.
5. Real-Time Systems Group. MARVIN – An autonomously operating flying robot. http://pdv.cs.tu-berlin.de/MARVIN. TU Berlin.
6. Marek Musial, Uwe Wolfgang Brandenburg, and Günter Hommel. Inexpensive system design: The flying robot MARVIN. In *Unmanned Air Vehicle Systems: Sixteenth International UAVs Conference*, pages 23.1–23.12, Bristol, UK, 2001. University of Bristol. ISBN 0-86292-517-7.
7. Volker Remuß, Marek Musial. Communication system for cooperative mobile robots using ad-hoc networks. In *Proc. of the 5th IFAC Symposium on Intelligent Autonomous Vehicles (IAV 2004)*, Lisboa, July 2004. Elsevier Science. ISBN 008-044237-4.
8. Volker Remuß, Marek Musial, and Uwe Wolfgang Brandeburg. BBCS – Robust communication system for distributed systems. In Thomas Christaller et al., editors, *Proc. IEEE Internat. Workshop on Safety, Security and Rescue Robotics (SSRR 2004)*, May 2004. ISBN 3-8167-6556-4.

Development of an Embedded Intelligent Flight Control System for the Autonomously Flying Unmanned Helicopter *Sky-Explorer*

Wang Geng[1], Sheng Huanye[1], Lu Tiansheng[2]

[1] Intelligent Computer-Human Interface
[2] Special Robot
SJTU-TUB Joint Institute for Information & Communication Technologies
Shanghai Jiao Tong University, China
{wgwtd,hysheng,tslu}@sjtu.edu.cn

Summary. The *Sky-Explorer* is designed to be an autonomously flying robot based on a model helicopter which can be used to take videos or to take some special missions in air. The *Sky-Explorer* system consists of a helicopter with an embedded computer system and necessary sensors on board, and a ground station which is a network of one or more Linux notebook PC. In order to achieve a relatively high-performance and long life formalproduct, we aim at the scalable, programmable, maintainable and portable onboard computer architecture. Then a function distributed computer structure was designed and adopted on board.

1 Introduction

The *Sky-Explorer* of Shanghai Jiaotong University (SJTU) is an unmanned small-scale mode helicopter. It was designed to be an autonomously flying robot in the future and can be used to take photos or videos in air, or to take some special missions autonomously.

Rotorcraft-based unmanned aerial vehicles (RUAV) have unique flight capabilities such as hover, vertical take-off/landing, pirouette, and sideslip, which cannot be achieved by conventional fixed-wing aircraft. These versatile flight modes make RUAV robotics to be a useful approach to perform tasks such as data and image acquisition of targets and affected areas, localization of targets, tracking, map building and others. And now, intelligent RUAV has been made possible through technological advances in various fields such as artificial intelligence, robotics, wireless communication, control theories and so on.

The *Sky-Explorer* of SJTU is developed based on the successful experiences of MARVIN which is an autonomously flying unmanned helicopter of Technology University of Berlin (TUB). MARVIN is an abbreviation for Multi-purpose Aerial Robot Vehicle with Intelligent Navigation. It is

Günter Hommel and Sheng Huanye (eds.), Embedded Systems Modeling – Technology, and Applications, 121–130.

an autonomously flying robot that can fulfill a complicated search mission purely on the basis of sensor data, and won the 2000 International Aerial Robotics Competition (IARC) Millennial Event.

At SJTU, aimed at the wider and wider utilization of intelligent RUAV in the future, an intelligent RUAV project has also been undertaken now. The ultimate objective of this project is to develop a high-performance and relatively long-life versatile formal RUAV product. Then the research work is mainly focused on to achieve a scalable, programmable, maintainable and portable on-board embedded computer control system. The scalability, programmability and portability will make the system easy to be upgraded and maintained, and easy to be adapted to use on another similar helicopters, and thus make the system higher performance, more versatile, and then a longer life.

In order to achieve the characteristics of scalability, programmability, maintainability and portability of the RUAV control system, distributed computer architecture was adopted to manage and execute the missions on-board. This paper mainly presents the design of the on-board computer system of *Sky-Explorer* and briefly introduces the overall system.

2 System Design

Figure 1 provides an overview of the *Sky-Explorer* system. Taking into account the scalability, programmability, portability, reliability, maintainability of the system, special attention was given to the hardware and software architecture of the on-board computer.

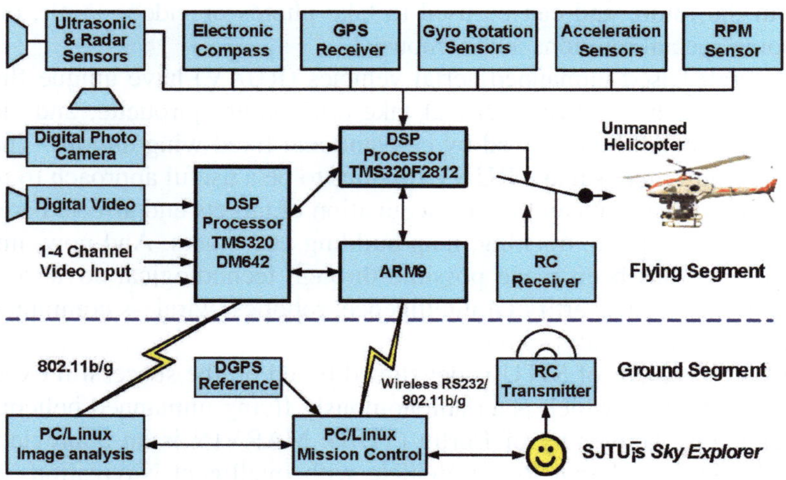

Fig. 1. Overview of the SJTU *Sky-Explorer* system

2.1 Air Vehicle

The basis of the *Sky-Explorer* air vehicle is a conventional model helicopter which is a kind of class 60 product of HIROBO. It has a rotor diameter of 153cm and a longitudinal fuselage length 138 cm, and is equipped with a two-stroke blended fuel engine. Its usable payload is about 4 kg.

Its controls are driven by five control actuators using PWM signal:

- Two servos for cyclical pitch adjustment of the swashplate of Hiller flap, then to control the pitch and roll direction of main rotor, that is front, rear, left and right.
- one servo for collective torque control of main rotor.
- one servo for tail rotor pitch control.
- one servo for engine throttle control.

2.2 Sensors

Just as Fig. 1 has shown, based on GPS receiver, the on-board computer control system gets the position and speed parameters in world coordinates. By three dimensions gyroscopes, the control system gets the helicopter rotation angular velocities and then numerically integrates them into corresponding angels. Actually, some other auxiliary sensors are necessarily been used on board to fulfill the control work or to promote the control quality. By three dimensions acceleration sensors, the control system gets the virtual lateral, longitudinal and upright accelerations which are used to stabilize the helicopter or to control the helicopter to proper flying attitude. By electronic compass, the yaw and pitch angles of the earth's magnetic field vector within the helicopter system are measured and are used to control the flying orientation or to correct the accumulated integration error of gyroscopes at certain conditions.

Other sensors are used to ensure the flying is safe. The echo-sounder using ultrasonic or radar transducer are used to measure the altitude over ground and to find the obstacles ahead of the flying route. A light barrier RPM sensor is used to measure the main rotor rotation speed. In many applications, a digital photo camera is necessary to accomplish the task, and sometimes even a video device is necessary to take some special task such as the real-time reconnaissance or visual navigation by real-time image analysis. The photo and video solutions is described in the following subsection.

2.3 On-board Computer System

The core of the whole intelligent control system is the on-board computer system, and the foundation of a high performance computer system is its powerful hardware system. In order to make the on-board computer system powerful enough to match the continuously increasing software requirements, distributed computer architecture is adapted on *Sky-Explorer*. Just as Fig. 1 has presented, the on-board distributed computer system is designed to be composed of three different processors. The first is a DSP processor TMS320F2812, the second is another DSP processor TMS320DM642 and the third is an ARM9 microprocessor. Every processor has its special outstanding features. Fig. 2 depicts the distributed structure of the three microprocessors.

The TMS320F2812 is a new powerful DSP processor of TI. In C2000 generation it is the highest performance 32-bit DSP-based controllers up to now. It is a highly integrated, high performance solution for demanding control applications. The C28x core is the world's highest performance control optimized core and offers of the computational bandwidth to handle numerous sophisticated control algorithms in real-time. Here are some of the attractive characteristics of TMS320F2812 in this application:

- High-performance 32-bit CPU up to 150 MIPS.
- 16 channels 12-bit ADC at up to 12.5 MSPS fast conversion rate.
- Up to 6 independent PWM waveforms (outputs) can be generated simultaneously.
- Up to 6 Capture-Unit can be used to measure the servo control PWM signals at no CPU load.
- 128K $\times$16 Flash and 18K $\times$ 16 SRAM on chip, and up to 1M$\times$1 6 external memory.
- Peripheral Interrupt Expansion (PIE) block that supports 45 peripheral interrupts.

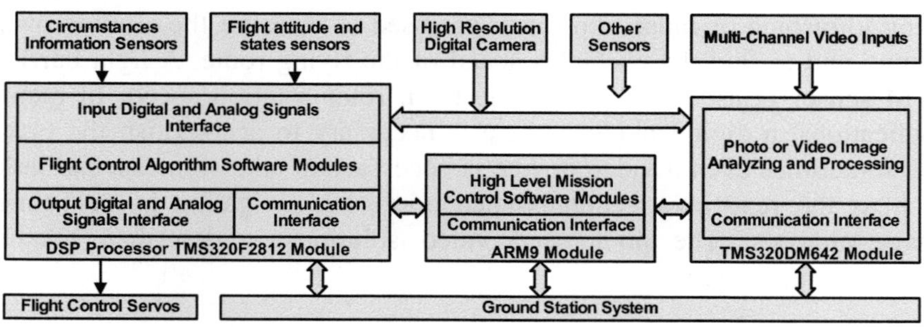

Fig. 2. Overview of the on-board distribute computer structure

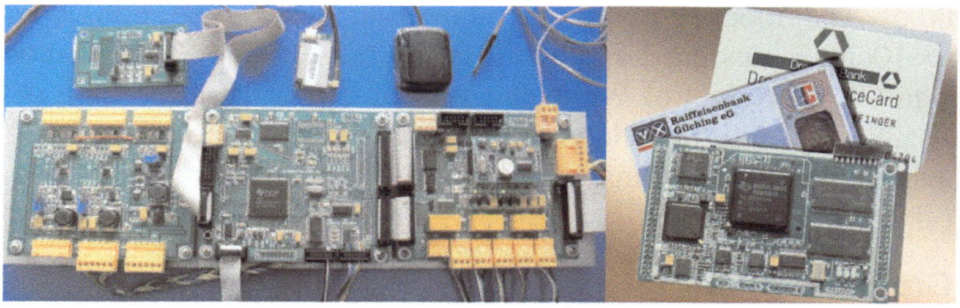

Fig. 3. The F2812 module and DM642 OEM Module

With on-board Flash memory, abundant peripheral interfaces and performance up to 150 MIPS, the F2812 is very suitable for this complex flight control of RUAV. Those all-around features ensure that almost all the sensors and servos can be interfaced with only one chip of F2812. And its high operation speed ensures it is powerful enough to run the sophisticated control algorithms easily.

The TMS320DM642 is another high performance DSP processor of TI. The TMS320C64x™ DSPs (including the TMS320DM642 device) are the highest-performance fixed-point DSP generation in the TMS320C6000™ DSP platform. With performance of up to 5760 MIPS at a clock rate of 720 MHz, The DM642 device has three configurable video port peripherals, and each video port consists of two channels. These video port peripherals provide a glueless interface to common video decoder and encoder devices, and support multiple resolutions and video standards.

Thus, based on this high-performance digital media processor DM642, up to six digital video cameras can be equipped on the unmanned helicopter. For example, 4 video cameras can be fixed at front, left, right, and downward direction of the helicopter respectively, and then to monitor the four directions simultaneously in real time. Based on the powerful DSP operation capability, the decoded video data can be compressed efficiently on chip in real time to proper video compressed format to reduce the storage space and transmission band. In advanced applications, the synchronous images from two special arranged cameras can be used to get 3D image which can be used to do more significative work.

The ARM9 architecture microprocessor used in this on-board distributed computer system is mainly to deal with the high level mission control task. Beside the 32-bit high performance and low power consume advantages, because several kinds of Embedded Operation System (OS) can be embedded in the ARM9 architecture microprocessor, many software resources can be used to speed the developing process. Linux OS and ANSI C program language are used in developing the ARM9 module.

Parts of The module board are shown in Fig.3. The F2812 board (12cm × 10cm) is designed and achieved by ourself. The DM642 module now is an OEM product of Wentech Corp. of China. Its dimension is just like a bank card. In the following work, the DM642 board and ARM9 board will be designed and integrated on board by ourself.

2.4 Ground Station

The basic ground station computer is a notebook PC running Linux Operation System. The DGPS reference receiver while needed is connected to the host via a RS232 interface. This computer serves as the user interface to the *Sky-Explorer* system: It accepts the flying mission, allows giving the "start" and "stop" commands, and displays the mission progress. Throughout the system development, the ground computer provided valuable help in the real-time monitoring of arbitrary parameters. The operation performance can easily be scaled by using more than one computers connected via TCP/IP protocol.

2.5 Communication

Data communication between *Sky-Explorer* and the ground station is performed via a pairs of wireless RS232 data communication modules whose reliable communication distance is at lease 500m outdoors at 38400 baud rate.

The second link is the digital communication signal of the conventional remote control transmitter Futaba-T9ZHP, which serves two purposes: ***a.*** it allows flying the *Sky-Explorer* manually for test purposes or in an emergency, and manual control is switched on or off via the RC system. ***b.*** It serves as technically independent trigger for necessary emergency termination mechanism.

When transmit the compressed video data, a broad-band wireless net

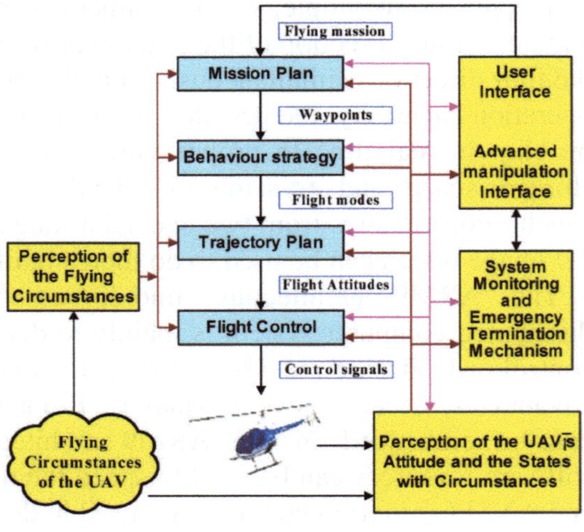

Fig. 4. The 4 levers hierarchical control structure of *Sky-Explorer*

transmission module based on protocol 802.11b/g is used on board. When exextend the WLAN on ground by scaled more AP (Access Point), the flying area with valid communication is extended also.

3 System Software structure

Like a man to perform the control operations, the auto mission system of *Sky-Explorer* is performed by a hierarchical control system. Figure 4 shows th 4 levers hierarchy structure, and the modularized software structure is shown in fig. 5.

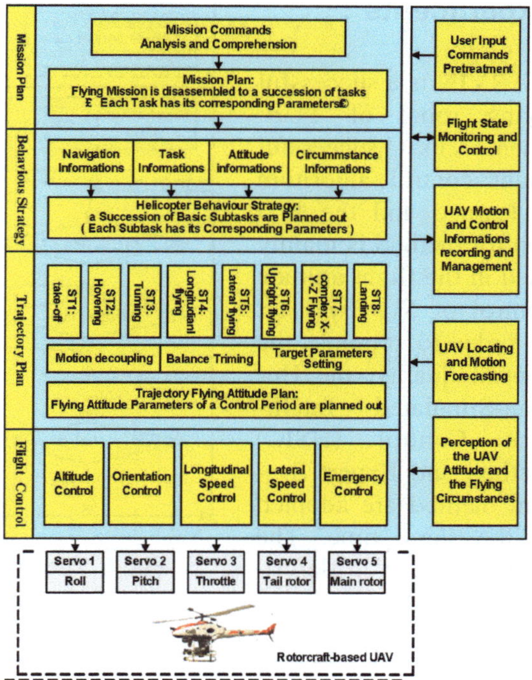

Fig. 5. Main modularized components of the mission control software

Based on the powerful performance of the onboard distributed computer system, those function modules in Fig. 5 are distributed efficiently into the three microprocessors according to their different characteristics just as depicted in fig. 2. The F2812 module mainly devotes to do the low lever real-time flight control work. The execution mode of the flight control modules is shown in Fig. 6.

The highest priority 32-bit timer0 Interrupt Service Routine (ISR) is used to schedule and manage the whole real-time tasks. The shared data blackboard in fig. 6 is a shared memory block which includes all the necessary parameters shared by all the software modules. The ISRs tell the foreground main loop routine what to do by setting the corresponding flag variables based on the analysis of current states on the shared data blackboard. The flight control module is actuated at a rate of 50Hz by the timer0 ISR and is called from the foreground main loop. Together with a maximum run cycle time for each task, the fulfillment of real-time requirements can be easily guaranteed by this approach at no computational overload.

4 Flight Control Algorithm and Experiments

MARVIN is a successful flying robot of TUB. Then, its successful flight control algorithm is learned and the PID control method is mainly used in the flight attitude control of *Sky-Explorer*. In order to promote the control quality, the Human-Simulated Intelligent Control (HSIC) method and Expert System method are adopted at the same time. The HSIC method is described detailedly in reference [5].

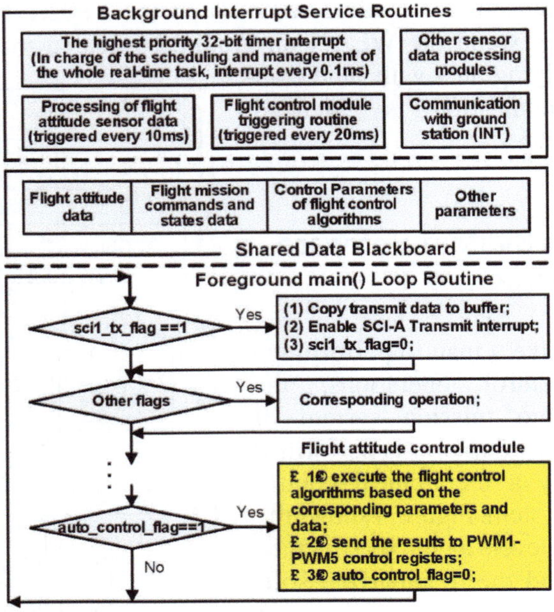

Fig. 6. The execution mode of the flight control module

Figure 7 shows the *Sky-Explorer* which was being used to record the pilot's remote control data by the on-board F2812 control computer. From the pilot's manual control data, the flight control rules can be derived and are used to construct the flight control Expert System. In order to make the flight control experiments safe and efficient enough at first developing period, a set of 6 degrees of freedom (DOF) ground test bedframe was designed and was used to test the flight attitude control algorithms. In those experiments, the on-board computer system works well. The detailed introduction of the flight control algorithm and the corresponding experiments of *Sky-Explorer* will be described in the next papers.

Fig. 7. Recording the pilot's control data by the on-board computer

5 Conclusion and Acknowledgements

Based on the on-board distributed embedded computer system described detailedly in this paper, the complex flying mission control work is distributed into three different powerful macro processors. This structure and design achieved the high-performance and scalability requirement of the hardware of *Sky-Explorer*. Any processor's upgrade will increase the system performance with a comparatively low cost on hardware and software modification. Each processor's abundant and powerful peripheral interface ensured the flexibility of the system, and then ensured the programmability of the system. The modularization of the control software and the inherent advantage of the Expert System, which is the independence of the Knowledge Base and the Reasoning Machine, can offer a certain possibility of transferring the control system to other similar helicopters. Then the system may be a long-life developing platform for RUAV.

During the development of the *Sky-Explorer* system, warmhearted helps were continuously supported from TUB. Here we wish to present our sincere thanks to the research group of Prof. Günter Hommel at TUB. Thanks very much for their warm interviews and advices about UAV technologies.

References

1. Sun Zhenping (2004) An Intelligent Control System for Autonomous Land Vehicle. Ph.D. thesis, National University of Defense Technology, China
2. H. Jin Kim, David H. Shim (2003) A flight control system for aerial robots: algorithms and experiments. Control Engineering Practice 11: pp 1389-1400, Elsevier
3. H. Jin Kim, David H. Shim, Shaiikar Sastry (2002) Flying robots: modeling, control and decision making, Proceedings of the 2002 IEEE International Conference on Robotics &Automation. Washington, DC. vol. 1: pp 66-71
4. J.V.R. Prasad, A.J. Calise, Y. Pei, J.E. Corban (1999) Adaptive Nonlinear Controller Synthesis and Flight Test Evaluation on an Unmanned Helicopter. Proceeding of the 1999 IEEE International Conference on Control Applications, Kohala Coast-Island of Hawai'I, pp 137-142
5. Li Zushu, Tu Yaqing (2003) Human-Simulated Intelligent Control. National Defence Industry Press, Beijing
6. Marek Musial, UweWolfgang Brandenburg, Günter Hommel (2000) Development of a flight control algorithm for the flying robot MARVIN. 2000 TUB-SJTU Joint Workshop on Advanced Robotics and its Application. Shanghai Jiaotong University

7. Marek Musial, UweWolfgang Brandenburg, Günter Hommel (2000) Cooperative autonomous mission planning and execution for the flying robot MARVIN. Intelligent Autonomous Systems 6: pp 636-643, IOS Press, Amsterdam
8. Nakanishi H, Inoue K (2003) Development of autonomous flight control system for unmanned helicopter by use of neural networks, Neural Networks 2003, Proceedings of the International Joint Conference on, vol. 3 pp 2400-2405
9. S. Kannan, C. Restrepo, I. Yavrucuk, L. Wills, D. Schrage, J.V.R. Prasad (1999) Control algorithm and flight simulation integration using the open control platform for unmanned aerial vehicles, Digital Avionics Systems Conference, 1999. Proceedings. 18th, vol. 2 pp 6.A.3.1-6.A.3.10
10. U. W. Brandenburg, M. Musial, G. Hommel (1998) MARVIN: Technology University Berlin's never-depressed flying robot for the IARC 98. In Proc. AUVS Symposium, Washington D.C., USA
11. Stuart Russell, Peter Norvig (2003) Artificial Intelligence: a Modern Approach(Second Edition), Pearson Education Press
12. Volker Remuß, Marek Musial, U. W. Brandenburg (2004) BBCS - Robust communication system for distributed systems. In Proc. IEEE International Workshop on Safety, Security and Rescue Robotics SSRR 2004. Bonn (Germany)
13. MARVIN: Multi-purpose Aerial Robot with Intelligent Navigation. http://pdv.cs.tu-berlin.de/MARVIN/, TU-Berlin

Model Predictive Control with Application to a Small-Scale Unmanned Helicopter

Du Jianfu, Lu Tiansheng, Zhang Yaou, Zhao Zhigang, Wang Geng

Special Robot
SJTU-TUB Joint Institute for Information & Communication Technologies
Shanghai Jiao Tong University, China
{djf2717,tslu,yaou_zhang,zhzg,wgwtd}@sjtu.edu.cn

Summary. Unmanned Helicopters are hard to control because of their nonlinear dynamic model and the constraints on actuators and outputs. Generally, PID, LQR, NN etc. is hard to handle the constraints problem. This paper presents a novel control method, model predictive control, which can deal with the input and output saturation constraints well for unmanned helicopters. The final results show that model predictive controller works well on the control of roll and pitch axes. It reduces the settling time and maximum overshoot and performs good stability and robustness comparing with PID controller.

1 Introduction

Small-scale unmanned helicopters are widely used in military and commercial field, because of its superior flight characteristics: vertical take-off and landing, hovering, longitude flight, lateral flight, pirouette, and bank to turn. Possible applications include close-up inspection of power lines, terrain surveying, search-and-rescue missions, filming movies, and the investigation and clean-up of hazardous waste sites.

Many universities, such as Technical University of Berlin, Georgia Institute of Technology, Stanford, UC Berkley, Waterloo, etc. are researching on unmanned helicopters. But the helicopter is a multi-variable, cross-coupling controlled target and is hard to control. From 2002, our team began studying the control of a small-scale model helicopter as show in Fig. 1.

Fig. 1. Small-scale helicopter

Currently, many control methods, such as PID, adaptive nonlinear control, neural networks control, fuzzy control, are used in helicopter controller design. But all these methods have a common disadvantage. They have

Günter Hommel and Sheng Huanye (eds.), Embedded Systems Modeling – Technology, and Applications, 131–139.

to first measure the output and compare with the desired output, then generate the control signal, so the control input always lags with the output. Comparing with these methods, model predictive control (MPC) generates the control signal based on the desired output not only on the output error. MPC can get satisfying control precision even the model is very rough (Yugeng Xi 1993).

Jonas Balderud (2002) and Arkadiusz S (2003) once used MPC controller to control a 2 DOF toy helicopter. In this paper, DMC control algorithm is introduced to control a 6 DOF small-scale helicopter and compare with PID control algorithm.

2 DMC Principles With Constraints

DMC includes three parts: model based prediction, feedback modifying, move optimization (Yugeng Xi 1993, Wu Fuming et al. 1994).

The model is used to predict the outputs according to the information from the past and the future input signal. Feedback modifying is used to decrease the influence of model error and load disturbance.

The control strategy of DMC is to minimize a performance index:

$$\min J(k) = \left\|w_p(k) - \tilde{y}_{PM}(k)\right\|_Q^2 + \left\|\Delta u_M(k)\right\|_R^2 \tag{1}$$

$$C\Delta u_M(k) \ll l \tag{2}$$

Where,

$Q=\mathrm{diag}(q_1 \cdots q_P)$, $R=\mathrm{diag}(r_1 \cdots r_M)$,

Δu_M is a vector of changes of the control signals,

P and M denote optimization and control horizons, respectively,

q_i and r_i are penalty coefficients.

This is a quadratic programming (QP) probem. How to solver the QP problem is the key to the optimization control. Considering the control performance and real time characteristic, we use optimization algorithm under constraints based on matrix decomposes technology.

Firstly, we solve the optimized $\Delta u_M(k)$ without constraints. It can be computed as:

$$\frac{dJ(k)}{d\Delta u_M(k)} = 0 \tag{3}$$

$$\Delta u_M(k) = (A^T QA + R)^{-1} A^T Q[w_P(k) - \tilde{y}_{P0}(k)] \tag{4}$$

Then check the constraints conditions (2) and separate the expressions that don't meet the conditions, and let them meet the following condition.

$$C_1 \Delta u_M(k) = l_1 \tag{5}$$

Where, C_1 and l_1 are the elements that don't meet the conditions from C and l.

So, the performance index is modified as:

$$\min J'(k) = J(k) + \|C_1 \Delta u_M(k) - l_1\|_S^2 \tag{6}$$

Where, $S = sI$, s is a very large positive number.

Then,

$$\Delta u'_M(k) = (A^T QA + R + C_1^T SC_1)^{-1}[A^T Q(w_P(k) - \tilde{y}_{P0}(k)) + C_1^T Sl_1] \tag{7}$$

Using the matrix decomposes technology and letting $s \to \infty$, we can get

$$\Delta u'_M(k) = \Delta u_M(k) - PC_1^T (C_1 PC_1^T)^{-1}[C_1 \Delta u_M(k) - l_1] \tag{8}$$

Where, $P = (A^T QA + R)^{-1}$

P can be computed off-line, only $(C_1 PC_1^T)^{-1}$ whose dimensions having been decreased needs computed on-line. This only need short computation time.

3 Controller Design

3.1 Helicopter Model

Helicopter model can be got through model revising, structure analyzing, system identification and neural networks (Chen Haosheng and Chen Darong 2003). System identification is a good modeling method to avoid the complexity involved with detailed modeling.

Daigo Fujiwara (2004) proposed a simple black-box system identification method and obtained single-input/single-output (SISO) non-cross coupling stable models on a Hirobo SF40 helicopter. Cross-validation results showed close agreement in the respective output signals obtained by simulation and by experiment.

The transfer function was derived as follows, for the $\dot{\theta}$ axis

$$\frac{\dot{\theta}}{u_\theta} = \frac{166.4}{s^3 + 14.30s^2 + 157.3s + 1331} \tag{9}$$

For the $\dot{\phi}$ axis:

$$\frac{\dot{\phi}}{u_\phi} = \frac{655.4}{s^3 + 19.20s^2 + 307.2s + 4096} \tag{10}$$

Our helicopter has similar structure with Hirobo SF4, so we can use their mathematics model to design controller and analysis it's performance.

3.2 MPC Controller

The helicopter control system consists of two levels: high level and low level. The low level controller consists of inner-loop angle velocity control, mid-loop attitude control and outer-loop velocity control. The high level generates the position and velocity command. The paper mainly discusses the application of MPC controller on the angle velocity and angle of pitch and roll axes. The control structure of the pitch axis is showed in the Fig. 2. The roll axis control structure is similar to the pitch axis.

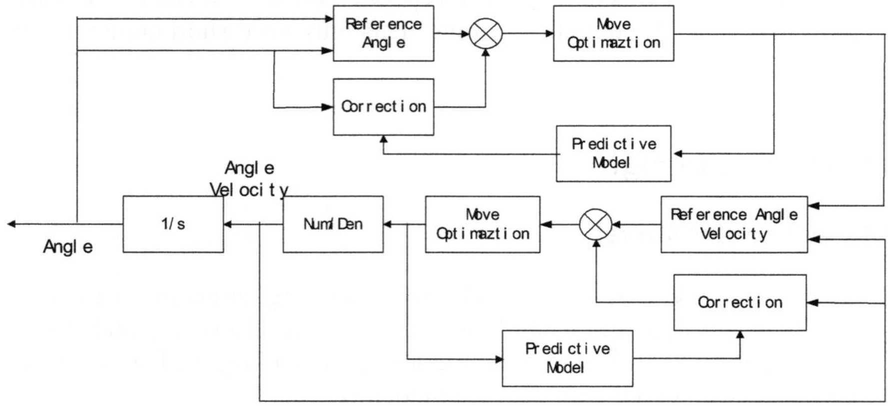

Fig. 2. Control structure of the pitch channel using MPC controller

Figure 3, Fig. 4, Fig. 5 and Fig. 6 are unit step responses of angle velocity control of pitch axis and roll axis, angle control of pitch axis and roll axis, respectively.

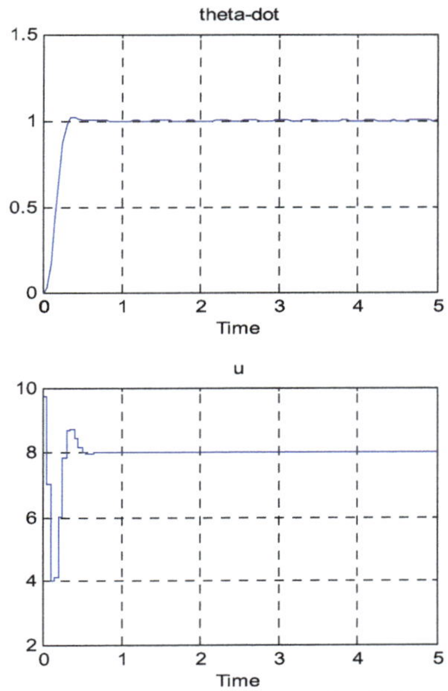

Fig. 3. Step response of MPC controller for $\dot{\theta}$

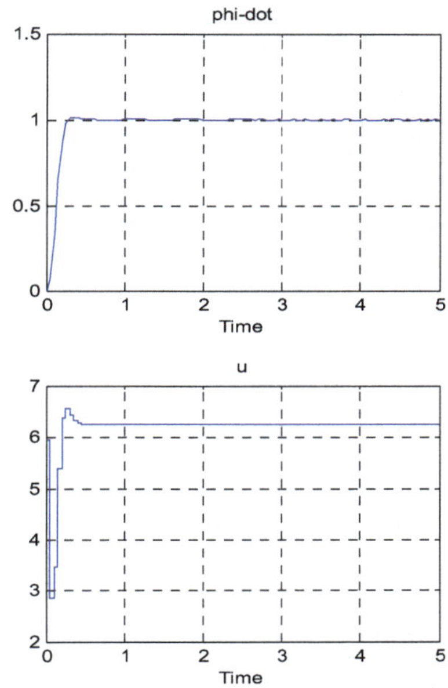

Fig. 4. Step response of MPC controller for $\dot{\phi}$

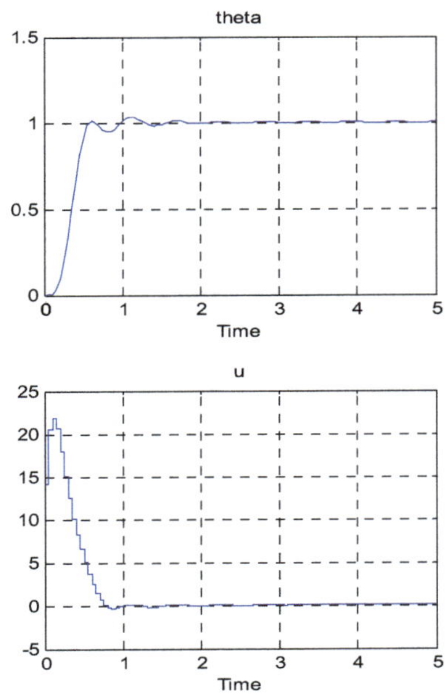

Fig. 5. Step response of MPC controller for θ

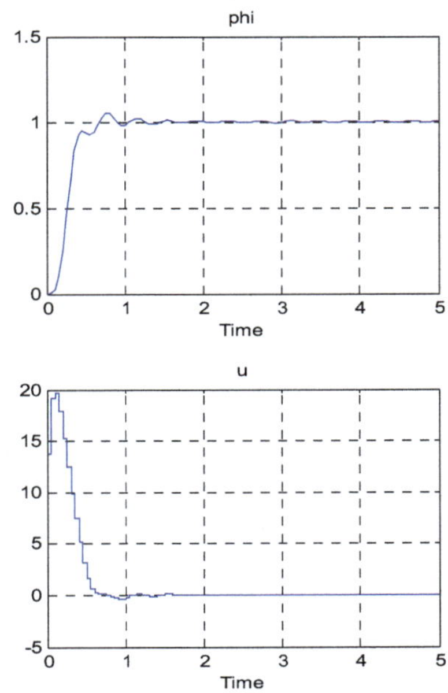

Fig. 6. Step response of MPC controller for ϕ

The sampling time is 0.05s, control horizon is M, optimization horizon is P, error weighting matrix $Q = diag(ones(P,1))$, Control weighting matrix $R = diag(0.05*ones(M,1))$, input constraint $U_{min}=-180$, $U_{max}=180$.The individual parameters are shown in table 1.

Table 1. Control parameters and performance index of MPC controller

	M	P	Overshoot[%]	Settling time[s]	Energy Consume[u^2]
A	2	10	1.8	0.3	6385
B	2	10	1	0.25	3883
C	3	12	3	0.55	2500
D	3	12	5	0.75	1842

A: Angle velocity control for $\dot{\theta}$, C: Angle control for θ

B: Angle velocity control for $\dot{\phi}$, D: Angle control for ϕ

3.3 PID Controller

PID controller is similar with the MPC controller. The control structure of pitch axis is shown in Fig. 7.

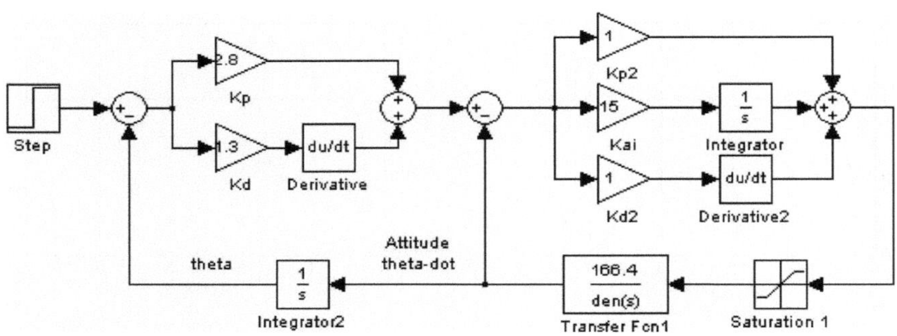

Fig. 7. Control structure of the pitch axis using PID controller

Figure 8, Fig. 9, Fig. 10 and Fig. 11 are unit step responses of angle velocity control of pitch axis and roll axis, angle control of pitch axis and roll axis, respectively.

The sampling time $Ts = 0.05s$, $U_{min} = -180$, $U_{max} = 180$. The individual parameters are shown in table 2.

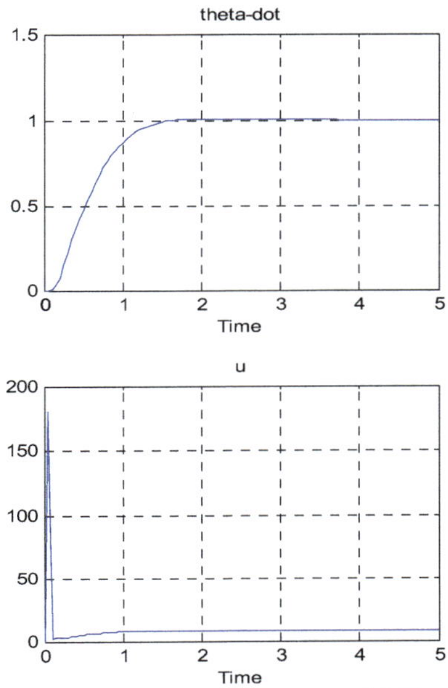

Fig. 8. Step response of PID controller for $\dot{\theta}$

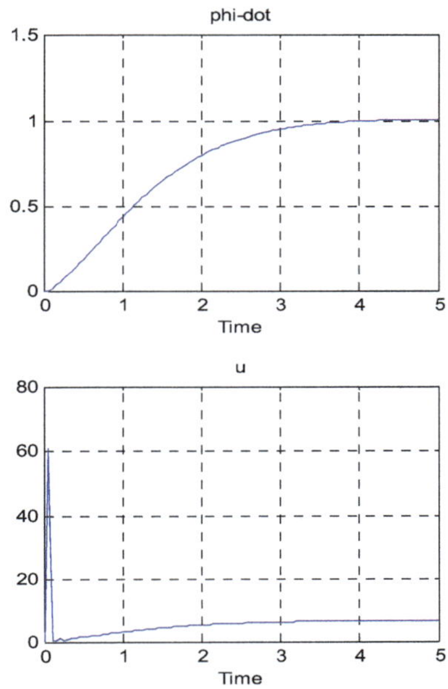

Fig. 9. Step response of PID controller for $\dot{\phi}$

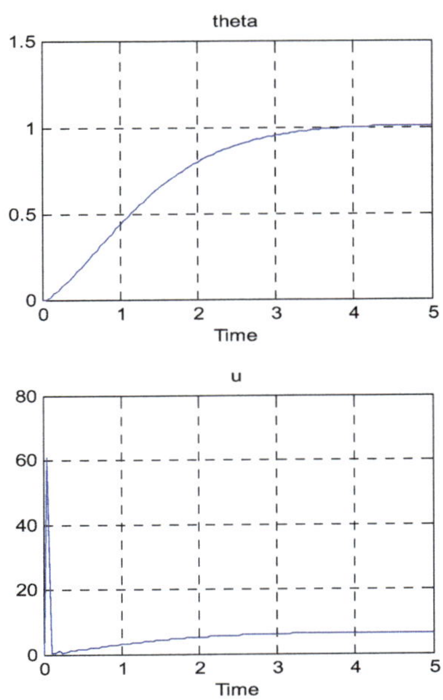

Fig. 10. Step response of PID controller for θ

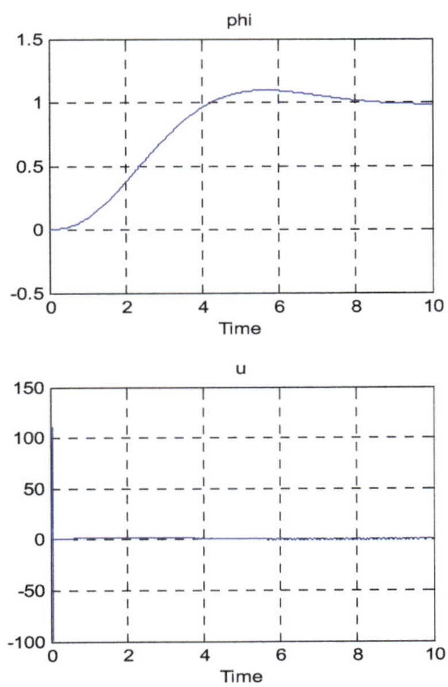

Fig. 11. Step response of PID controller for ϕ

Table 2. Control parameters and performance index of PID controller

	Kp	Ki	Kd	Overshoot [%]	Settling time [s]	Energy Consume [u^2]
A	1	15	1	0.5	1.25	37993
B	1	5	3	0.4	3	6311
C	2.8	0	1.3	8	2.15	33560
D	0.7	0	0.5	9	4	23685

A: Angle velocity control for $\dot{\theta}$, C: Angle control for θ

B: Angle velocity control for $\dot{\phi}$, D: Angle control for ϕ

3.4 Comparing Results

The Performance of MPC and PID controller is shown in table 3.

Table 3. MPC controller and PID controller performance

	MPC			PID		
	Overshoot [%]	Settling time [s]	Energy Consume[u^2]	Overshoot [%]	Settling time [s]	Energy Consume [u^2]
$\dot{\theta}$	1.8	0.3	6385	0.5	1.25	37933
$\dot{\phi}$	1	0.25	3883	0.4	3	6311
θ	3	0.55	2500	8	2.15	33560
ϕ	5	0.75	1842	9	4	23865

The table shows that MPC and PID controller both have small overshoot, but MPC controller has shorter settling time and smaller energy consume. For example, the settling time of unit step responses of roll velocity controlled by PID controller is 12 times that of MPC controller. This is because the MPC controller has prediction function and input energy consume is considered when design.

4 Hardware Implementation of the Control System

From the simulation results, we can see that MPC controller has better performance than PID controller. The controll algorithm can run on our onboard computer.

The onboard computer is a Digital Signal Processor, TMS320F2812, which is highly integrated, high-performance solution for demanding control applications. The high-performance 32-Bit CPU with harvard bus architecture runs at 150 MHz. It has 16 channels 12-bit ADC, motor control peripherals, one SPI and two UARTs .The controller is suitable to handle the mathematics computation and can implement complicated control algorithms. It is convenient to use and competent for the flight control.

5 Conclusions

The paper present two controllers, MPC and PID, for flight control. The simulation results shows that MPC controller has better performance. MPC controller has shorter settling time and smaller energy consume than PID controller. On the other hand, MPC controller needs more computation time and consumes lots of resources of the processor. Decreasing the computation time will be our next research direction.

References

Arkadiusz S. Dutka, Andrzej W. Ordys, Michael J. Grimble (2003). Non-linear Predictive Control of 2DOF helicopter model. Proceedings of the 42nd IEEE Conference on Decision and Control, Mad, Hawaii USA, pp. 3954-3959.

Chen Haosheng, Chen Darong (2003). Current Researches on the Micro-Helicopter's Dynamics Modeling. Flight Dynamics. 21(3):1-5.

Daigo Fujiwara, Kenzo Nonami (2004). H∞ hovering and guidance control for autonomous small-scale unmanned helicopter. Proceedings of 2004 IEEE/RSJ International Conference on Intelligent Robots and Systems. Sendai, Japan, pp. 2463-2468.

Jonas Balderud and David I. Wilon (2002). Application of predictive cotrol to a toy helicopter. Proceeding of the 2002 IEEE International Conference on control Applications. Glasgow, Scotland, U.K pp. 1225-1229.

Wu Fuming, Sun Demin, Zheng Jingbo (1994). An Improved DMC Algorithm used in Measure Failure and it's Application on Paper Machine Control. Proceedings of the IEEE International Conference, pp. 148-152.

Yugeng Xi (1993). Predictive control, Press of China National Defense Industry, Beijing.

The Recurrent Neural Network Model and Control of an Unmanned Helicopter

Zhang Yaou, Lu Tiansheng, Du Jianfu, Zhao Zhigang, and Wang Geng

Special Robot
SJTU-TUB Joint Institute for Information & Communication Technologies
Shanghai Jiao Tong University, China
{yaou_zhang,tslu,djf2717,zhzg,wgwtd}@sjtu.edu.cn

Summary. A complete method of modeling and control of small-scale helicopter is presented. In order to get the model of the helicopter, an experiment platform of the unmanned helicopter is set up. On the experiment bench, firstly control the helicopter approximately in hover, and then maintain the main rotor pitch angle and tail rotor pitch angle invariable, and next change the longitudinal cyclic control and lateral cyclic control of the helicopter to obtain the corresponding longitudinal and lateral attitude angle signal and then record corresponding input output data. Secondly, the attitude model of helicopter is proposed. The recurrent neural network is used to identify the proposed model structure. The accuracy and the validity of the model are proved by the simulation result. Finally, neural network controller is used to control the attitude nonlinear system, then the quality of the control results are tested by the simulation.

1 Introduction

With the development of science and technology, unmanned small scale helicopters have been expected by many fields such as agricultural spraying, atmospheric monitoring, surveillance, target acquisition. For these missions, an autonomous flight control of the helicopter is indispensable. The autonomous flight control requires many technologies such as obstacle avoidance as well as attitude and position control.

Now, the existing small model-scale helicopter realizes only a small part of the helicopter's inherent qualities. For example, their operation is limited to hover and slow-speed flight; their control performance is sluggish. Although most multivariable control methods, such as LQR, LTR, EA method, are available now, yet an accurate model should be formulated first before using these control method, while dynamic models for a particular model-scale helicopter are not easily to get. Thus most UAVs still used classical control systems such as single-loop PD systems. And their controller parameters are usually tuned manually for a distinct operating point (Martin

Günter Hommel and Sheng Huanye (eds.), Embedded Systems Modeling – Technology, and Applications, 141–148.

and Hanns 1994; Alejandro and Rogelio 2004), the difficulty of modeling prohibits from getting the good flying quality of the helicopter.

There are many methods in modeling the small-scale helicopter. Using a standard first-principle based modeling approach, considerable knowledge about rotorcraft flight dynamics is required to obtain the governing equations (Konstantin and Carsten 2004; Erik 1999). The modeling method based on system identification has been developed and successfully used with model-scale helicopters. But experiment design and the data processing were not provided clearly in the article (Bernard Mettler et al., 1999). The articles (Fahmida 2003; Guozhi Ma and Saleh 1998; K.S. Narendra and K. Parthasarathy 1990) give the neural network model of yaw movement or vertical movement. This method is very useful in controller design in the reality. This article is to use neural network to model the longitudinal and lateral attitude of the small-scale helicopter on the platform of our own. With the model, the controller can be designed to fulfill the autonomous flight of the helicopter.

The paper is arranged as follows: Section 2 describes the principle and structure of the model-scale helicopter; a detail description of the construction of the bench is given in section 3; in section 4, some aspects and details of the neural network modeling are presented, and the model is validated by the simulation result. In section 5, the control method is design; the control parameter is achieved based on the neural network. Section 6 contains a summary of the experience with the test bench and the method of the modeling and controller design and gives an outlook to the next work in the future.

2 Helicopter

The helicopter used in the our project in Shanghai Jiao Tong University (SJTU) is a commercially available model-scale helicopter powered by the engine, with two-bladed main rotor of 1.5m diameter and a pair of stabilizer blade and a tail rotor. The relatively rigid blades are connected to a yoke through individual flapping hinges. The yoke itself is attached to the rotor shaft over a teetering hinge. The Bell-Hiller stabilizer bar consists of two small rotor blades mounted at the right angle to the main blades on the rotor shaft and linked to those mechanically (Bernard Mettler et al. 1999). It receives the same cyclic control input as the main rotor. The stabilizer bar flapping motion is used to generate a control augmentation to the main rotor. It is not used to produce lift and it reacts slowly to the control inputs, pitch and roll disturbance than the main blades and less sensitive to airspeed and wind gusts.

This model scale helicopter is controlled by four inputs. They are: collective pitch input (θ_c); collective tail rotor pitch input (θ_T); longitudinal cyclic stick input (B_{1s}); and lateral cyclic stick input (A_{1s}), these four inputs control the states of the helicopter through the five servos. The collective pitch changes the angle of attack of the main rotor blades independently of their location on the rotor disk. This produces the vertical force, which causes the helicopter to climb or descend. The collective pitch angle of the tail rotor controls the yaw motion. With the longitudinal (B_{1s}) and the lateral (A_{1s}) inputs, the angles of attack of the main rotor blades are changed sinusoidally depending on their location on the rotor disk. With these cyclic inputs, on the one hand, pitch angle and forward position, on the other hand, roll angle and sideward position, are controlled.

Fig. 1. The Vertical Platform

3 Description of the Frame

Our team aims at controlling the flight of a small helicopter in a three-dimensional space. However, the model helicopter system is a complicated system, in order to gain insight into the modeling of the helicopter, the aerodynamic effects, the ground effect and the sensors, we have built a flying stand that allows the helicopter to move in the vertical axis, to turn around the vertical axis. The stand also has joints, which allow rotation around the longitudinal and lateral axes as indicated, these joints of the pitch and roll axes were restricted, limiting angle ranges of the roll and the pitch of the helicopter. The vertical platform is depicted in Figure 1.

In order to find out the coupling between the pitch and the roll, the movement of the height and the joints of the yaw are clamped, the roll and pitch angles are obtained through a standard angular encoder. The data pairs of the input control PWM signal and the output angles were got using data acquisition cards (PC-7483/7426).

The motivation for mounting the helicopter on a stand as opposed to flying untethered is twofold. First, by measuring the joint angles of the stand, an inexpensive and accurate means for measuring the attitude of the helicopter is obtained. Second, it provides a safe setup for performing experiments in laboratory.

4 The Method of Neural Network Modeling

The objective of the identification experiments is to get pitch and roll model of the helicopter near hover. Because the helicopter is a strong cross couplings nonlinear state-space MIMO system, neural network modeling methods are preferred. Ideally we would like to be able to drive the helicopter with a sufficiently rich input to excite all the dominant modes and cross-couplings. There are two major obstacles. First, the helicopter is open loop unstable, so a stabilizing trim command is necessary when taking data. Second, the helicopter is nonlinear so we can only expect to get reasonable results when we use small-signal excitations around the trim position.

4.1 Data Collection

There are many parameters that affect the attitude of the helicopter, such as the speed of the rotor, collective pitch angle, collective tail rotor pitch angle, longitudinal cyclic input; and lateral cyclic input. If the helicopter flies freely, either of this parameter change will influence the motion of the pitch and roll attitude of the helicopter. In our experiment, the speed of main rotor is kept 1550r.m.p. The speed of the main rotor and the collective pitch angle are controlled by the same stick, so the collective pitch angle is also kept constant. The collective tail rotor pitch angle and the speed of the tail rotor are kept constant. It is only the longitudinal cyclic stick input and lateral cyclic stick input that the pilot can adjust. On one hand, these two values will be sent to control helicopter, on the other hand they will be gathered by the computer. Thus the pitch and the roll angles of the helicopter and the control signals will be got.

The procedures for the experiments are as follows:

1. Set the input variable value of main rotor speed at a fixed value and gradually change the variable value of tail blade angle and the longitudinal cyclic input and the lateral cyclic input to make the helicopter approximately at hover.
2. Keep the rotor speed and the collective pitch angle and the tail blade angle constant, and gradually change the stick input of the lateral stick and the longitudinal stick simultaneity, but the angles of the pitch and roll of the helicopter are not too far form the trim position.
3. Repeat the step of 1 and 2 ten times and record the input output pairs, the sample frequency is 50hz.

Parts of the experimental data are shown in Figure 2. On this figure, x-axis presents the sample point numbers, the curve Power is the pitch angle in figure 2a, and roll angle in figure 2b. All the data have been filtered before enter the computer.

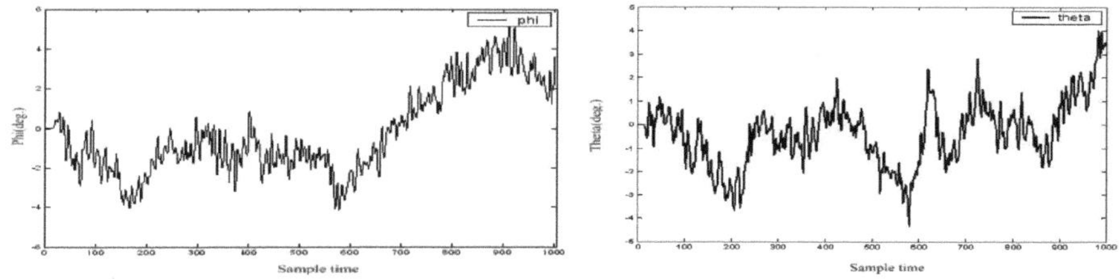

Fig. 2. The pitch and roll angle Experiment data of small-scale helicopter

4.2 Neural Network Modeling

Neural networks can successfully model nonlinear dynamic systems which are often described by non-linear difference equations, especially the discrete time input and output models (Lingji Chen and Narendra 2002; K.S. Narendra and K. Parthasarathy 1990;Guozhi Ma and Saleh 1998), such as

$$y_{k+1} = f(y_k, y_{k-1} \cdots y_{k-n+1}, u_k, u_{k-1} \cdots u_{k-n-1}) \tag{1}$$

Where $u \in U$ is the input variable, $y \in Y$ is the output variable, and f is a real analytic function defined on $U \times Y$. For the neural network has the ability of good function approximation, feed forward neural networks with $2n$ inputs and one output can be used to model the function f. These neural networks have $2n$ input vectors, which consist of n past inputs and n past outputs of the dynamic system.

Equation (1) can be represented as (2), Where d is the delay of the system.

$$\begin{aligned} y(k+d) &= \sum_{i}^{n} (b_i \phi(\omega_{i1} y(k) + \cdots + \omega_{in} y(k-n-1) + \omega_{i,n+1} u(k) + \cdots + \omega_{i,2n} u(k-n-1))) \\ &= B^T \phi(W X_{K-1}) \end{aligned} \tag{2}$$

The perception neural network (PNN), with one hidden layer with sigmoid nonlinear function and one output layer with linear function, is adopted. The network is trained using incremental BP. Training the NN implies estimating the matrices W and B, knowing the input and output data of the dynamic system and assuming suitable function of $f(\cdot)$.

According to the theory of the helicopter and Newton-Euler principle, the order the function is approximately 3. Also from the data collection, we can find that the current helicopter pitch and roll angles are determined not only by the current values, but also by the previous three sample time values of them. If we select three steps delay, the next pitch and roll can be described with the following function, in which the coupling of the pitch and roll is taken into consideration. The speed and the pitch angle of main rotor and those of tail rotor are all kept constant, so it is not considered in the neural network. The dynamics of pitch and roll are shown in the equations (3), (4).

$$\Phi(k) = f_1(\Phi(k-1), \Phi(k-2), \Phi(k-3), Ua, Ub, \Theta(k-1), \Theta(k-2), \Theta(k-3)) \quad (3)$$

$$\Theta(k) = f_2(\Phi(k-1), \Phi(k-2), \Phi(k-3), Ua, Ub, \Theta(k-1), \Theta(k-2), \Theta(k-3)) \quad (4)$$

A multi-layered perception NN was used to model this plant. It was trained using incremental BP. This neural network has three layers, the input layer, the hidden layer, and the output layer, the layers of this network possesses 8,15,2, respectively. The inputs to the network are $\Phi(k-1), \Phi(k-2), \Phi(k-3), Ua$, $Ub, \Theta(k-1), \Theta(k-2), \Theta(k-3)$. The outputs are $\Phi(k)$, $\Theta(k)$. The first nine experiments data have been used to train the neural network and the tenth experiment data is used to simulate and validate the model. The comparison of pitch angle of the helicopter between the tenth experiment results and the simulation results by the neural network model is shown in Figure 3a .The solid line in the figure means the experiment data, while the dashed data means the simulation data. It can be seen that the NN can almost approximate all the experiment data except the plant output at several negative or positive peaks. The comparison of roll angle of the helicopter between the tenth experiment result and the simulation result by the neural network model is described in Figure 3b. The solid line in the figure means the experiment data, while the dashed data means the simulation data. It is almost the same as that of the roll angle. From the two figures, we can see the error between the experiment and the prediction is very small and can be neglected. So this method can be fit to predict the movement of the attitude angle in reality.

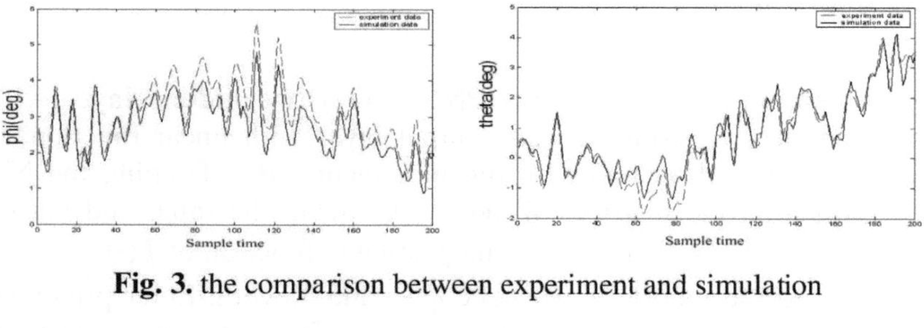

Fig. 3. the comparison between experiment and simulation

5 Controller Design and Implementation

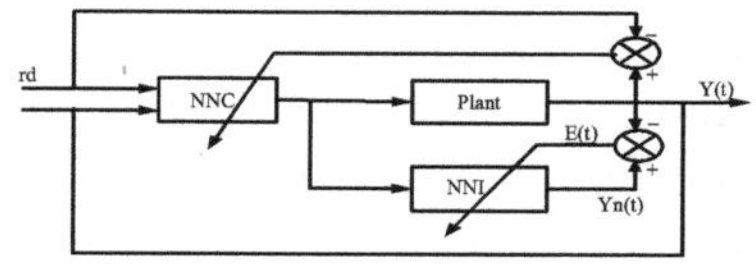

Fig. 4. Neural Network Control Structure.

The objective of this study is to design a controller to provide decoupled tracking of pitch and roll angles to make the actuators of overall closed-loop system acceptable for flight control. The control structure can be seen in figure 4 where the controller can be design based on the model in section 4. The inputs to the neural controller are actually outputs of the plant and the desired outputs value, and the structure of the neural controller is multi feed forward neural network with 4-10-2 in three layers respectively. The training algorithm is the incremental Bp algorithm. The method is validated by the simulation. The angles of the pitch and roll were originally set to be 5 degree and 6 degree respectively, and reach approximately zero very quickly with the controller, the control results are shown in figure 5.

TMS320F2812 is a high integrated, high-performance 32-bit CPU especially fit for control application. It has many features that we need in our control system, especially its supporting real-time mode operation and C++ programming. So TMS320F2812 is chosen to process our control codes and tested algorithm in the experiment bench.

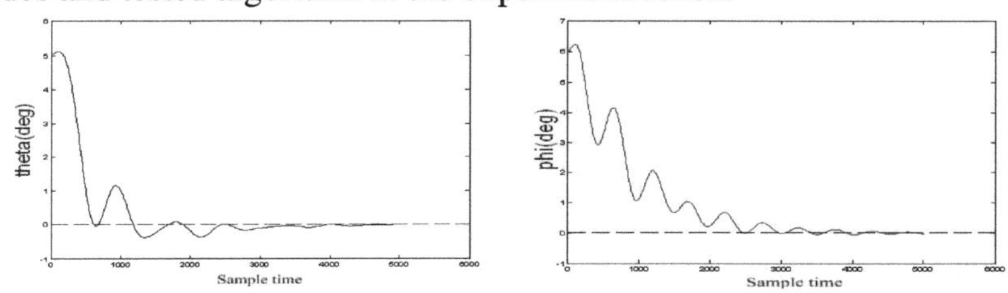

Fig. 5. Control Effect

6 Conclusion and Outlook

An accurate method of modeling the unmanned helicopter using the recurrent neural network method and a neural network controller are presented in this paper. First, the test bench is designed to gather the attitude and control data of the flying helicopter. Secondly, neural network method is used to obtain the model of the system. Finally, based on the model established, the neural network controller is designed to fulfill the stable control.

The helicopter is mounted on the test bench, the velocities of the helicopter in x, y, z directions and the yaw movement are all considered zero, but the transition velocities have great influence on the angles of pitch and roll of the helicopter, thus the model and the stable controller in the bench are not as those are in the real flying. The model and the control method should be modified before used in the real sky exploring. The next work is to take the velocity and the yaw movement of the helicopter into consideration to get the full model with this method and design the corresponding controller.

References

Alejandro Dzul, Rogelio Lozano (2004) adaptive control for radio-controlled helicopter in vertical flying stand, international journal of adaptive control and signal processing, Int.J.Adaptive Control signal process. 18:473-485

Bernard Mettler and Takeo Kanade ,Mark B. Tischler(1999)System Identification Modeling of a Model-Scale Helicopter, Annual Forum Proceedings - American Helicopter Society, v 2, 1706-1717

Erik Skarman (1999) A helicopter model, Linköping Electronic Articles in Computer and Information Science Vol. 4:nr 14

Fahmida N.Chowdhury (2003) A new approach to real-time training of dynamic neural networks, international journal of adaptive control and signal processing, Int. J. Adapt. Control signal process. 17:509-521

Guozhi MA, Saleh Zein-Sabatto(1998) Intelligent flight control design for helicopter yaw control, IEEE

Lingji Chen, Kumpati Narendra (2002) Intelligent control using neural network and multiple models, proceedings of the 41st IEEE conference on decision and control Las Vegas, Nevada USA, December

Konstantin Kondak and Carsten Deeg (2004) Mechanical model and control of an autonomous small size helicopter with stiff main rotor, proceedings of Int. conference on intelligent robots and systems, Sep. 23-Oct. 2, sandai, Japan

K.S. Narendra and K. Parthasarathy (1990) Identification and control of dynamical system using neural networks, IEEE Trans. on neural networks, 01.1, No.1, 4-27

Martin F. Weilenmann, Hanns P. Geering (1994) A test bench for rotorcraft hover control, Journal of Guidance, Control, and Dynamics, v17, n4, Jul-Aug, 729-736

GA-Based Evolutionary Identification of Model Structure for Small-scale Robot Helicopter System

Zhao Zhigang and Lu Tiansheng

Special Robot
SJTU-TUB Joint Institute for Information & Communication Technologies
Shanghai Jiao Tong University, China
{zhzg,tslu}@sjtu.edu.cn

Summary. This paper analyses problems in the model identification and express the transfer characteristics of small-scale robot helicopter using the fractional form. Subsequently, a GA-based identification method is used to identify it. Our proposed method is verified through an example and an experimental study. The example can see that the genetic algorithm used to identify the transfer function is useful and feasible. The experimental results reveal the robot helicopter system structure and the identification of parameters can be simultaneously attained using the proposed GA method, in contrast to the conventional identification algorithms.

1 Introduction

It is known that helicopters can take off and land vertically, perform hover flight as well as cruise flight, and have agility and maneuverability, so they are used in many areas. In recent years, because its high cost-effective, security and characteristics mentioned above, small-scale unmanned helicopters have attracted many interesting of experts[1]. The model must be attained no matter what we want to research some items about helicopter, such as controllability, stability and so on[1].

The traditional modeling approaches is the first principle approaches, which requires comprehensive knowledge of helicopter as well as extensive flight experimentation. The most effective and accurate approach to obtain linear small-scale helicopter models is through linear system identification. Theory researches and practice applications indicate that traditional model identification algorithms require systems have good error and noise statistic characteristics. There is large error coming forth in the helicopter modeling process, the candidate model structures often be omitted or improper[2]. If the model structure is not correct, an accuracy result can't be gained by using conventional or advanced parameter estimated methods.

Günter Hommel and Sheng Huanye (eds.), Embedded Systems Modeling – Technology, and Applications, 149–158.

It is shown by simulation research that the genetic algorithm-based identification method is of good practicability, satisfactory results can be got with it, no matter what kind of input signal is used. There are many complicated system, such as robot helicopter and mechatronic systems, which are difficult to model using traditional modeling approaches. Many novel evolutionary algorithm using genetic algorithm can autonomously determine the optimal models[3,4,5,6]. Aiming at the complicated structure, nonlinearities and high frequency operated of robot helicopter, this paper presents an evolutionary algorithm for the identification of robot helicopter systems. In this research, under the assumption that the transfer characteristics of the robot helicopter system can be modeled by a fractional expression, the GA assigns the optimal combination of modules for poles and zeros, and identifies those parameters to satisfy the specified system transfer characteristic, by evaluating the appropriate fitness in the genetic process. The proposed scheme has a distinctive feature in that the determination of the robot helicopter system structure and the identification of parameters can be simultaneously attained, in contrast to the conventional identification algorithms, e.g., a linear regression techniques under the given system structures.

The remainder of this paper is organized as follows: Section 2 discuss about the expression of robot helicopter system model. Section 3 use an example to exploit that GA-based identification method has many advantage comparing with conventional identification methods. We can interpret the results of GA-based identification method using an experiment as in the following sections. Section 5 gives the conclusions and future directions of research. Finally, the authors give sincerely acknowledgments to the people and organize which give useful helps to our works.

2 The Expression of Robot Helicopter System Model

The main objective of the research is to identify the structure and parameters of small-scale helicopter in one degree random, based on a transfer function, so we must to express the robot helicopter system model with a model set. The rotor is used to produce the lifting force as well as the forces and torques that are used to control the rotorcraft. Unlike the wing of a fixed-wing aircraft, the rotor is a dynamic system[4], so the dynamic of the helicopter must include the rotor and the stabilizer bar dynamic. In this paper we develop the dynamical model of robot helicopter in one random degree in hover flight condition, we can simplify the motion dynamic of robot helicopter with SISO linear transfer functions.

The pilot operate the helicopter in hover condition, subsequently, only one small-signal excitation is used to excite the dynamic characteristics of one degree of freedom, and at same time, other input controls is adjusted continually to keep the helicopter steady[7,8]. In the following, the roll degree of freedom is used as an example to express the process of modeling the helicopter's model structure.

The roll motion of helicopter is related to the flapping motion of the rotor, i.e. the relation between the flapping angle β and the lateral cyclic input δ_ϕ, the relation between the flapping angle β and the roll moment M_ϕ are must be ascertained, then the dynamic characteristics of roll degree of freedom will be understood and the parameter model structure will be attained.

Making reference to the flapping motion expression of a full-scale helicopter,

$$\ddot{\beta} - \cos\xi\ddot{\phi} + 2\Omega\sin\xi\dot{\phi} - \sin^2\xi\beta\dot{\phi}^2 + \Omega^2\beta = \frac{n\Omega^2}{8}\frac{\rho ac_2\left(B^4R_2^4 - R_1^4\right)}{I_f}\left(\frac{\alpha_1L_7L_8}{L_5L_6L_9}\delta_\phi\sin\xi - \frac{\alpha_2\left(\dot{\beta} - \dot{\phi}\cos\xi\right)}{\Omega}\right) \tag{1}$$

and making a reasonable simplicity, we obtain the useful and simple results as follow:

$$\beta_{\max} \approx \frac{\alpha_1P_2}{\alpha_2}\delta_\phi - \frac{2\Omega}{\alpha_2P_1}\dot{\phi} \tag{2}$$

$$\xi_{\max} \approx \pi \tag{3}$$

where, $\alpha_1, \alpha_2, P_1, P_2$ and Ω are constants, δ_ϕ and ϕ are variables.

As same we make reference to a full-scale helicopter, the expression of the roll moment is defined as follow:

$$M_\phi = \frac{n\rho\Omega^2acR^4B^4}{16L_1(L_2+L_3)}\left(\alpha_3\left(\frac{L_3L_8}{L_9}\delta_\phi - L_2L_4\beta_{\max}\cos\xi_{\max}\right) - \alpha_4L_1(L_2+L_3)\frac{\dot{\phi}}{\Omega}\right) \tag{4}$$

We substitute $\beta_{\max}$ and $\xi_{\max}$ into the expression for M_ϕ from(4),

$$M_\phi = \left(\frac{\alpha_3P_3L_3L_8}{L_9} + \frac{\alpha_1\alpha_3}{\alpha_2}P_2P_3L_2L_4\right)\delta_\phi - \left(\frac{2\alpha_3P_3L_2L_4\Omega}{\alpha_2P_1} + \frac{\alpha_4P_3L_1(L_2+L_3)}{\Omega}\right)\dot{\phi} \tag{5}$$

Now we substitute M_ϕ into the Euler equation describing SISO roll dynamics, a simplified linear transfer function is obtained.

$$\frac{\phi(s)}{\delta_\phi(s)} = \frac{\left(\frac{\alpha_3P_3L_3L_8}{L_9} + \frac{\alpha_1\alpha_3}{\alpha_2}P_2P_3L_2L_4\right)\frac{1}{I_{xx}}}{s\left(s + \left(\frac{2\alpha_3P_3L_2L_4\Omega}{\alpha_2P_1} + \frac{\alpha_4P_3L_1(L_2+L_3)}{\Omega}\right)\frac{1}{I_{xx}}\right)} \tag{6}$$

Based on a transfer characteristic of the robot helicopter in one random mentioned above, the transfer function can be generally formulated by the polynomial form or the fractional expression. In the polynomial form, it is difficult to set a reasonable searching area for the parameters. If the area is too small, the optimal point can't be included; if the area is too large, the searching of genetic algorithm will be inefficient. Another reason is that the value of coefficients may not match. Many kinds of coefficients combination are insignificant, this will reduce the calculation speed[4]. So the SISO transfer functions of robot helicopter can be written as following forms.

$$G(s) = \frac{\omega(s)}{\delta(s)} = K \frac{\Pi_b (s + B_b)}{s \Pi_a (s + A_a)} \tag{7}$$

In this research, in order to express a high-order proper transfer function as a combination of simple factors, i.e. gain, poles, and zeros are utilized. By factorizing the transfer function of (7), it can be handled as the combination of factor modules as follows:

$$G(s) = \frac{\omega(s)}{\delta(s)} = K \frac{1}{s} \prod_p \left(\frac{1}{s + A_{1i}} \right) \prod_q \left(\frac{s + B_{2j}}{s + A_{2j}} \right) \tag{8}$$

In developing the model above-mentioned, we make many assumptions and simplifications, example, we consider the helicopter will flight in hover condition; we still have not modeled the helicopter engine and PWM servos; the assumptions and simplifications may be incorrect, etc., so it is necessary to ascertain the order of the transfer function using evolution methods.

3 Principle of System Identification Using GA

The basic concept of the GAs where developed by Holland and revised by Goldberg. Goldberg shows that the GAs are computationally simple and independent of any assumption about the search space. Actually they are stochastic parallel global search algorithms based on the mechanism of natural genetics and the biological theory of evolution. Because Gas exploit strategies of genetic information and survival of the fittest to guide their search, they need not calculate the gradient or assume that the search space is differentiable or continuous. GAs simultaneously evaluate many points in the parameter space, so they are more likely to converge toward a global solution. GAs are very suitable for searching discrete, noisy, multimodal and complex space. According to the characteristics of the robot small-scale helicopter, a kind of improved genetic algorithm is introduced into model identification.

Binary genes is a general coding method in simple genetic algorithm. It is used widely because of the simple coding, decoding, crossover, mutation operation. So it is used in this paper, though the string of binary genes is too long. The order, gain, poles, and zeros in (8) are expressed in binary chromosomes consisting of 3 bits each for p and q, 30 bits for K, and 20 bits for complex A and B with 10 bits each for real and imaginary parts. A set of the chromosomes composes an individual in the GA process as shown in Fig. 1. The genetic operations are individually process for each corresponding chromosome because the parameters in (8) are independent of each other. In the following experimental studies, a 276-bit individual (3-bit × 2+30-bit × 1+20-bit × 12) is designed to handle a model whose order is $k \leq 8 (p \leq 4, q \leq 4)$. This word length allows each individual in the population to express a mathematical model with arbitrary order, while only the part in A and B corresponding to p and q has effective information on the model [5].

p	q	K	A_{11}	…	A_{1i}	…	A_{21}	B_{21}	…	A_{2j}	A_{2j}	…

Fig. 1. Individual expression for parameters

Specifications of the GA operation are listed in Table 1, where guidelines to determine the genetic parameters are as follows:

- The size of the population should be selected considering the convegency under the practical computation load,
- The best individual will go forward to the next generation without undergoing any change,
- Crossover in a generation should be performed for almost every individual,
- Mutation rate should be 1%–3% to prevent the search from being a random one, with consideration of the population size and bit length of the individual.

Table 1. Specifications of GA operation

Population size	100
Tournament selection	
Multi-point crossover	crossover rate: 0.8
Bit mutation	mutation rate: 0.05

In the proposed identification, the variable parameters in the mathematical model are evolutionally identified by the genetic process so that the actual output and the model simulate input should coincide with each other. That is, the GA optimizes the combination of p, q, K, A and B which makes the fitness function F_{fit} in (9) minimum, using the time-series output response data $\omega_{M(n)}$ and $\omega_{Mm(n)}$ obtained from the plant system and the mathematical model, respectively

$$F_{fit}=\sum_{n=1}^{S}\{W_{f1}(\omega_{M}(n)-\omega_{Mm}(n))\}^{2}+W_{f2}k \tag{9}$$

where, S is the sampled number of the response data, $k=p+2q+3$ is the total number of structural parameters, and W_{f1}, W_{f2} are weights. In this fitness evaluation, the optimal identification for parameters, i.e., poles and zeros, can be attained under the minimum order of model structure since not only the integral of squared error but the order are evaluated considering the information criterion.

4 Example

In order to validate the realizablity of the proposed GA-based model identification used to attain the adequate fractional form, in first, a set of parameters is designated to form a transfer function like expression (2), we define it as follow function(10):

$$G_{e}(s)=\frac{(s-10-i20)(s-10+i20)}{s(s+23.5)(s+60.25+i30)(s+60.25-i30)(s+58.5+i3.2)(s+58.5-i3.2)} \tag{10}$$

Subsequently, a triangular wave for 2 seconds is given as an input signal and an output signal will be simulated. Input and output data are sampled with a sampling interval of 0.001sec, an example of the estimation data are shown in Fig. 2. We use these data to validate the method proposed in the chapter 3, the algorithm is convergent when it is up to 2000 generation. The resulted simulated parameters values and the real values (the parameters used in simulation) are shown in table 2. Using the comparability between simulated values and the real values, we can see that genetic algorithm used to identify the transfer function coefficients is useful and feasible.

Table 2. The results of the parameters in the example

Para.	p	q	A_{11}	A_{12}	A_{13}	A_{21}	A_{22}	B_{21}	B_{22}	K
Real	3	2	23.5	60.25 +i30	60.25 –i30	58.5 +i3.2	58.5 –i3.2	–10 +i20	–10 –i20	3.0 $\times10^{6}$
Simulate values	3	2	23.3 1	59.23+ i29.89	60.35– i30.57	58.36+ i3.24	58.57– i3.18	–10.38+ i20.15	–9.86– i19.92	2.97 $\times10^{6}$
Values with noise	3	2	23.2 9	59.19+ i29.78	58.92– i30.64	58.63+ i3.17	58.74– i3.29	–10.43+ i20.34	–9.67– i19.94	3.04 $\times10^{6}$

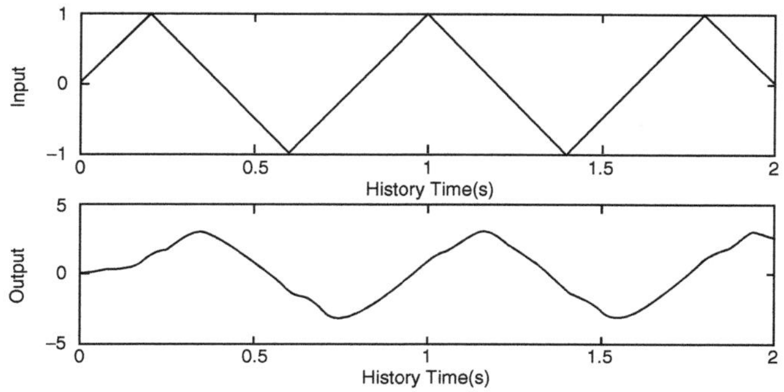

Fig. 2. Input and output data used in the example

To verify the performance of adaptive genetic algorithm applied in robot helicopter transfer function, white noise w_1 and w_2 of which amplitude are about 5% of the maximum of the input u and output y respectively are superposed on the primitive data, because in the actual application, the measure noise is inevitable. Using the mixed data, the simulated values are shown in table 2. The results express that the proposed GA can attain the exact parameters of the transfer function with the measure noise.

5 Experiment and Results

In the experiment, a kind of small-scale model helicopter was chosen as an experiment platform. Because of the small scale, the model helicopters are unstable and have a fast time-domain response. In order to overcome the shortcomings mentioned above, the flybar is designed to augment stability and make it easier for a pilot to fly. Because of the exist of the flybar, the modeling of the robot helicopter will be more difficult[7].

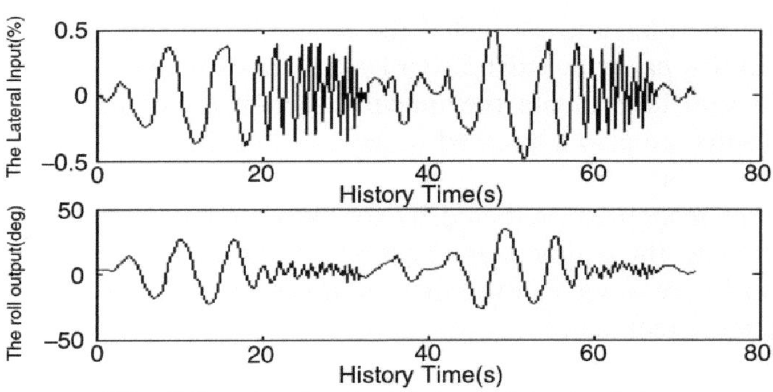

Fig. 3. Input and output data used to simulate

The helicopter is actuated by Futaba PWM servos, one each for lateral cyclic, longitudinal cyclic, main rotor blade pitch, tail rotor blade pitch, and engine throttle. Each servo is driven by a control signal of ± 0.5 ms modulated onto a 50Hz PWM waveform with a neutral position of 1.5ms. Pilot commands are provided through a six channel Airtronics RC transmitter.

In this experiment, attitude angles (ϕ, θ, ψ) and control signals ($\delta_\phi, \delta_\theta, \delta_\psi$) are measured. All data acquisition is done with the hardware and software developed by the members of Shanghai Jiaotong University's Autonomous Small-Scale Model Helicopter Project. Three angles sensors are added to measure the control signals. The datum is recorded for identifying three SISO pole-zero transfer function as (2). The datum contained the characteristics of the helicopter roll DOF is shown as Figure3.

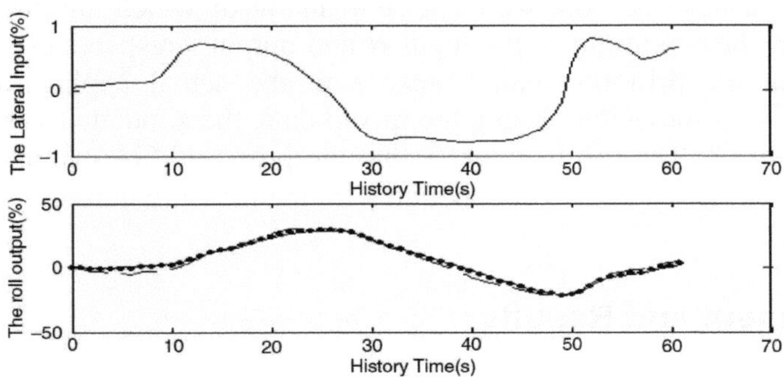

Fig. 4. Input and output data used to verify
The real value. The simulated value using LSQ. The simulated value using GA

Once the input-output data has been taken and stored for the isolated single DOF motion, the algorithm mentioned in section 3 is run to estimate the transfer function expressing the roll motion, the results are expressed in table 3. Sung K. Kim developed three transfer function of small-scale helicopter angle motions and estimated the coefficients using a least-squares method[1]. In this paper, we also use a least-squares method (LSQ) and our measured datum to estimate the model structure developed by Sung K. Kim, the results are also expressed in table 3. For the verification, another recorded inputs are used as inputs to the identified model to compare the helicopter responses predicted by the two models with the responses recorded during the flight test. The results from the comparison are presented in figure 4, we can see that the proposed GA-based method can produce better agreement.

Table 3. The results of the parameters in the experiment

Para.	p	q	A_{11}	A_{12}	A_{13}	A_{21}	A_{22}	B_{21}	B_{22}	K
LSQ	1	0	9.82	—	—	—	—	—	—	1765
GA	3	2	3.19	0.17	5.0	40.75 +32.3i	9.71	40.81 −32.5i	0.5	1.395

6 Conclusion

- This paper analyses problems in the model identification and express the transfer characteristics of small-scale robot helicopter using the fractional form.
- The proposed scheme has a distinctive feature in that the determination of the robot helicopter system structure and the identification of parameters can be simultaneously attained.
- The proposed method is not only applied in helicopter modeling, but also in some system that can not be measured exactly.

7 Acknowledgment

This research was supported by National Nature Science Fund (NNSF) of the people's republic of China, which afforded part of research fund. The members of Shanghai Jiaotong university's Autonomous Small-Scale Model Helicopter Project gave author helpful advice and constructive criticism. The authors would like to express sincere gratitude to them.

References

1. Sung K. Kim (2001) Modeling, identification, and trajectory planning for a model-scale helicopter. The thesis for the Degree of Doctor of Philosophy
2. Smith C B, Wereley N M (1998) Damping identification in helicopter rotor system. Annual Forum Proceeding. American Helicopter Society: pp 391–407
3. Ali Akramizadeh, Ali Akbar Farjami, Hamid Khaloozadeh (2002) Nonlinear Hammerstein model identification using genetic algorithm. Proceeding of the 2002 IEEE international conference on artifical intelligence systems
4. Liu Changliang, Liu Jizhen, Niu Yuguang, Yao Wanye (2002) The application of genetic algorithm in model identification. Proceeding of IEEE TENCON'02: pp 1261–1264
5. Makoto Iwasaki (2005) GA-Based evolutionary identification algorithm for unknown structured mechatronic system. IEEE Transactions on Electronics. VOL. 52, No. 1, 5: pp 300–305
6. Kazunari Inoue (2002) Identification of Continuous-Time Systems via Genetic Algoritom. SICE 2002.8: pp 1989–1993
7. Bernard Mettler (2001) Modeling small-scale unmanned rotorcraft for advanced flight control design. The thesis for the degree of doctor of philosophy
8. John C. Morris, Michiel van Nieuwstadt, Pascale Bendotti (1994) Identification and control of a model helicopter in hover. Proceedings of the Amaricas control conference: pp 1238–1241

Framework for Development and Test of Embedded Flight Control Software for Autonomous Small Size Helicopters

Markus Bernard, Konstantin Kondak, and Günter Hommel

Real-Time Systems and Robotics
Technische Universität Berlin, Germany
{berni,kondak,hommel}@cs.tu-berlin.de

1 Introduction

Over the last years, a many applications for small size Unmanned Aerial Vehicles (UAVs) were described. Especially reconnaissance missions can be carried out by these UAVs more efficiently compared with a full sized aircraft. As a platform for UAVs, model helicopters are quite attractive due to their maneuverability, large spectrum of commercially available vehicles and moderate costs. The first step of using UAVs is the establishment of safe autonomous flight capabilities. As it was shown in several publications e.g. [1], the implementation of robust control for small size helicopters requires real flight experiments. It is difficult to conduct such experiments (establishing of reproducible testing conditions, presence of an experienced safety pilot etc.) and these experiments are connected with high risk of helicopter loss.

In the presented paper, a framework for development and testing of embedded flight control software for autonomous helicopters is presented. This framework is composed of two main parts: the laboratory setup with small size model helicopter and the software system.
The framework allows to perform following tasks:

- development of a flight controller for a helicopter in Matlab/Simulink environment
- test the controller in simulation
- generate automatically embedded software for helicopter hardware
- test the controller in real flight experiments
- application of the generated embedded software on the end target for control and user interface

Günter Hommel and Sheng Huanye (eds.), Embedded Systems Modeling – Technology, and Applications, 159–168.

The mechanical part of the laboratory setup is crash resistant and allows testing developed control approaches with low risk of destroying the helicopter. In the presented laboratory setup, the helicopter performs a real flight without any reduction of existing degrees of freedom; different disturbances like wind gusts can be produced under well defined conditions. The framework is designed to be used with up to four cooperating autonomous helicopters.

2 System Overview

The laboratory setup, s. Fig. 1, is composed of an electric model helicopter LOGO-10 which is rigidly fixed in a safety cage, made of carbon tubes, and a measurement arm with 7DOF. All axes are unactuated and equipped with optical encoders. In Fig. 1 only the wrist joint of the measurement arm on the top of the safety cage is shown. The shoulder joint of the measurement arm is mounted on the ceiling.

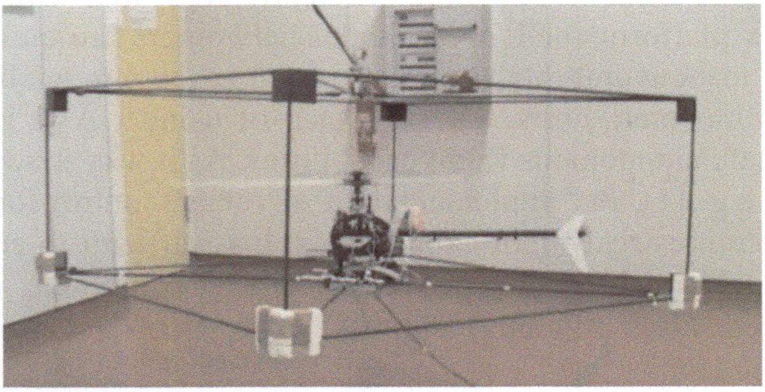

Fig. 1. Laboratory setup with helicopter and safety cage

This measurement arm is used only for the determination of the actual position and orientation of the helicopter and does not support it. Other systems for determining position and orientation of helicopters, e.g. vision based motion tracking systems, can be used with the framework. The total mass of the safety cage is about 1 kg.

The control hardware in the laboratory setup is composed of three main parts, s. Fig. 2:

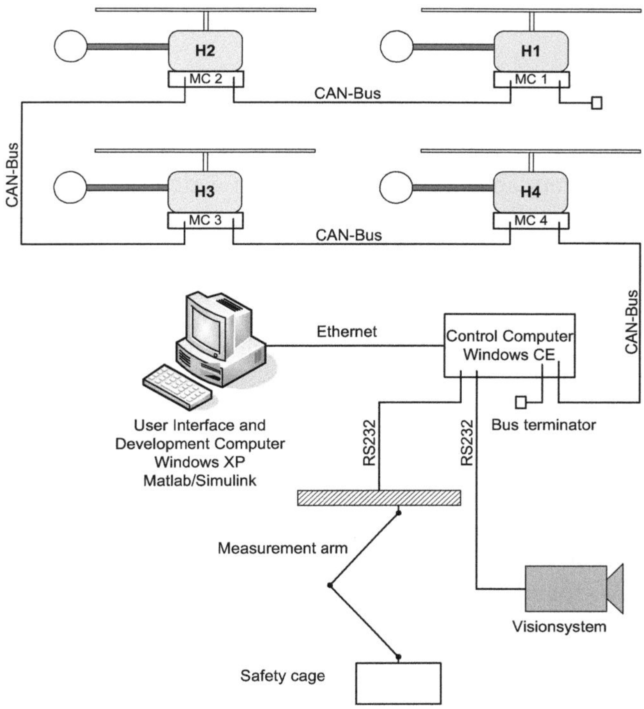

Fig. 2. System components

1. microcontroller board for controlling the helicopter actuators, for acquiring signals from sensors and from the inertial measurement unit as well as for interfacing with the control computer via CAN-Bus
2. control computer where the flight control software is running
3. user interface and development (UID) computer

By using multiple helicopters in the framework, all helicopters are connected to one CAN-Bus. The control computer can be mounted on one of the helicopters or be placed on the floor. The control computer is connected to the UID computer via TCP/IP-link.

The software system is the main part of the framework. The software system is composed of three parts:

1. Low level software for microcontroller board.
2. Real-time base system (RTBS) for flight control. RTBS runs under Windows CE and fulfils hard real-time requirements for flight control. The flight controller code is automatically generated from

Matlab/Simulink environment and automatically placed in the RTBS frame.

3. User interface and development (UID) software. This software is based on Matlab/Simulink environment where all inputs and outputs of the helicopters are accessible. The user is able to implement arbitrary controllers and user interfaces with a rich set of tools provided by Matlab/Simulink. The implemented controller can be tested first within the simulation environment before performing real flight experiments. After testing, the implemented controller can be automatically converted from Simulink model to C-code and placed in the RTBS flight controller frame using scripts incorporated in UID software (Simulink RTW is also required).

In sec. 3 the design concepts, structure and implementation of the main part of the software system - RTBS is explained in detail. Special attention is paid to the hard real-time requirements of the software system. Windows CE is not supported by Simulink RTW as target for code generation. The presented concept of RTBS allows using the generic code generator of Simulink RTW for almost each POSIX compatible operating system. In sec. 4 the real-time properties of the system were discussed and in sec. 5 the experimental performance evaluation of the RTBS is given. The conclusions are given in sec. 6.

3 The Real-Time Base System (RTBS)

Before the RTBS is explained, a short overview of the operating system should be given, since the real-time behavior of the RTBS depends on the underlying operating system (OS). Furthermore, the features of the operating system and its API[1] determine the complexity of the RTBS implementation.

The following hard real-time OS were compared [2]: RT-Linux, VxWorks, Windows CE and QNX. Windows CE was chosen as real-time OS, because it allows to switch easily between a PDA as real-time computer (e.g. for field tests) and a regular pc (e.g. for development) and its API is mostly compatible to Windows NT (2000, XP). Furthermore for Windows CE many drivers and a descent, low prized development environment are available. Most of the time dependent Windows CE functions (e.g. the *Sleep* function) exhibit a temporal resolution of one millisecond, which should be sufficient for the most applications. For high precision

[1] API = Application Programming Interface

time measurement Windows CE supports the *QueryPerformanceCounter* function, which allows to measure time with the maximum frequency supported by the underlying hardware timer. Windows CE features memory protection between different processes and therefore it is impossible to accidentally modify data that belong to other processes. All processes contain one or more threads with user defined priority. Different from Windows NT (2000, XP), Windows CE does not derive a basis threat priority from the underlying process. For thread prioritization Windows CE supports 256 different priority levels, but only the lowest 8 priorities a accessible via the *SetThreadPrioity* function and just the highest priority level (of this 8 levels) allows real-time operation. But if the used operation system image is self generated or the OEM[2] allows the usage of OEM restricted functions, it is possible to use the *CESetThreadPriority* function, which allows assigning 247 real time priorities. To preserve the possibility of switching between pc hardware and a PDA only the non OEM functions were used in the RTBS.

The RTBS was designed to meet the following requirements: modularity, standardized communication and synchronization between modules. The modules are modeled using Windows CE processes. Communication between the modules is realized using the blackboard concept. Synchronization is achieved with notification events. This blackboard architecture was realized as follows: Each module communicates with other modules using a DLL[3] called *hs_share*, which allows to access different data structures. These data structures are located in a shared memory buffer and only accessible using this DLL. Read accesses are decoupled from write accesses in a way that the last current data is returned, until a write access is completely finished. Write accesses on data structures are exclusive and a thread that requests write access will be blocked until a previous write access is finished. For each shared memory structure an event object is allocated by the DLL. This event is triggered automatically whenever a write access to that structure terminates. The *hs_share* DLL provides functions to wait for one or more events and to trigger events manually. Furthermore, it is possible to define additional events.

The scheme of the RTBS implementation is shown in Fig. 3. The whole system consists of four modules: *hs_kinematic*, *hs_communicator*, *hs_main* and *hs_matlab*.

---

[2]OEM = Original Equipment Manufacturer

[3]DLL = Dynamic Link Library

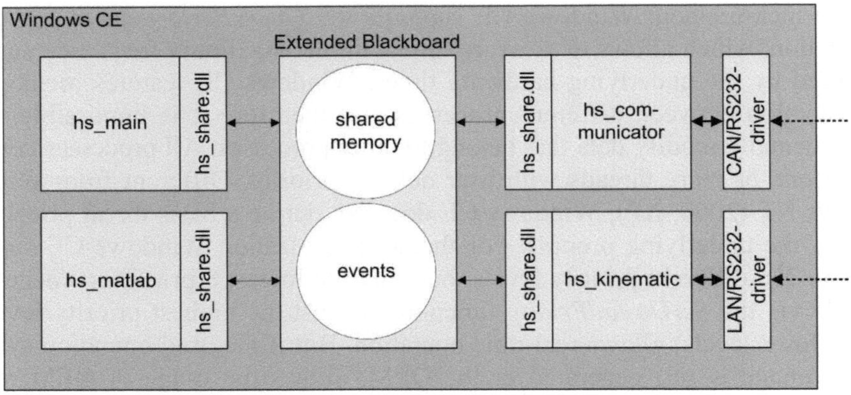

Fig. 3. The real-time base system

The *hs_kinematic* module determines the kinematical parameters (position and orientation) of the helicopters from the measurement arm or a vision system and provides the data to the blackboard.

The *Hs_communicator* was designed to handle the communication between the real-time computer and the micro controllers using a CAN-bus connection. The data throughput of the CAN-bus limits the communication to a maximum of four micro controllers. Additional to the CAN-bus the *hs_communicator* module supports an RS232 link for one micro controller. Furthermore, this module realizes global clock synchronization with the algorithm described in [3].

The purpose of the *hs_main* module is to launch, prioritize and stop the other modules. In addition, the module calls the initialize and the destroy functions of the *hs_share* DLL.

The *hs_matlab* module is of special importance, since it serves as a container for Matlab generated real-time code. Matlab generates this code from a Simulink model which implements the complete helicopter controller. The *hs_matlab* module itself is composed of three parts. First part is the invariant startup file, which contains the main function. In this file the *hs_share* DLL is set up and the Matlab main thread is created. Second part is the invariant Matlab libraries, required by the generated real-time code. The last part is represented by the generated real-time code itself. The *hs_matlab* calculation cycle runs as follows: New data arrives in the shared memory buffer of the blackboard. After that a notification event wakes *hs_matlab* and the data is copied into the Simulink model of the helicopter. A calculation step is performed and the result is copied from the Simulink model into the shared memory of the blackboard. Then the *hs_matlab* blocks itself until the next notification event for new data arrives. One part of the *hs_matlab* module that has not been mentioned above is the Matlab external mode interface. This interface is contained in the Matlab libraries, but the thread, required by the interface to work properly, is launched inside the startup file. The external mode interface allows connecting the real-time code to the Simulink interface. In this way it is possible to use the Simulink interface to start and stop the real-time code, change model parameters, view and log data. The external mode interface supports communication using RS232 or Ethernet, whereas the RTBS only supports the Ethernet connection.

4 Real-Time Properties of the System

Due to the design of the system all threads of the RTBS run periodic. Therefore, the results for rate monotonic scheduling [4] can be used to prove the real time properties of the system. For this the utilization bound should be calculated:

$$U(n) = n\left(2^{1/n} - 1\right)$$

where n is the number of all threads in the system: RTBS threads as well as Windows CE threads. It will be assumed that all Windows CE real-time threads run periodic and have a higher rate than RTBS threads. The threads of the RTBS have also lower priorities than the Windows CE threads. Due to the rate monotonic scheduling, the real-time properties of the system are satisfied if the maximal processor load is less than $U(n)$. For $n \to \infty$ is $U(n) \simeq 0.69$. The measured maximal processor load in the system (together with visualization and logging threads which are not real-time) is below 0.6. The threads of RTBS must have priorities proportional to their working rates: higher rate corresponds to higher priority. The round-robin scheduler of Windows CE has a huge time slot compared to the period of the considered threads. To that reason, the threads of RTBS with the same working rate could have the same priority.

5 Performance Evaluation of the RTBS

The real-time behavior of Windows CE was evaluated in many different publications [5-7]. For that reason the Windows CE operation system was not evaluated separately, instead the real-time behavior of the RTBS and Windows CE as a whole was considered. The RTBS itself was barely modified for the real-time tests. Only one module was added and the

RS232 connection of the *hs_communicator* module was used. The *MaxLoadCE* module contains a thread (of a certain priority) that generates the maximum possible load. Two different test series were accomplished, in the first series all measurements were performed in software and in the second series the outputs of the system were measured with an oscilloscope. For both series, the input data of the system is generated with a frequency of 100 Hz by an external micro controller and the real-time computer should generate result data with 100 Hz frequency. The software measurement was set up as follows: Every single measurement lasted 60 seconds, the first 10 seconds without additional load, and then the *MaxLoadCE* thread was started with a certain priority. After another 10 seconds, the thread terminated itself and the last 40 seconds the system ran without the additional load again. This setup made it possible to evaluate the unstressed system, the system under stress and how the system recovers after the stress period. The results of these tests are summed up in the following table[4]:

MaxLoadCE	$\Delta t_{received}$ (ms)			Δt_{send} (ms)			Δt_{Matlab} (ms)		
priority	min	max	mean	min	max	mean	min	max	mean
normal	9	11	10	9	11	10	1	2	1.54
highest	9	11	10	9	11	10	1	2	1.54
real-time	0	115	10	9	634	13.27	-	-	-

The time $\Delta t_{received}$ is the measured time between two received incoming data packets, Δt_{send} is the measured time between two outgoing packets and Δt_{Matlab} is the time needed for the Matlab calculations. It is easy to see that a thread below real-time priority does not affect the real-time threads. In case of *MaxLoadCE* at real-time priority $\Delta t_{received}$ (min) is 0[5]. This is explainable by the fact, that *MaxLoadCE* terminates after 10 seconds but the measurement continues. After *MaxLoadCE* has terminated, all received and buffered data packets are written into the shared memory. The maximum of $\Delta t_{received}$ is 115 ms, because the Windows CE scheduler interrupts *MaxLoadCE* after 100 ms and reschedules the threads, using the Round-Robin method. The remaining 15 ms are a displacement, caused by the frequency (100 Hz) of the arriving data and the calculation time of the remaining threads (after being delayed 100 ms). The minimum time of Δt_{send} is 9 ms, in contrast to the minimum 0 ms of $\Delta t_{received}$. The explanation is, that the send thread of *hs_communicator* depends on the results of the Matlab calculations and the used Simulink model was limited to 100 Hz.

[4] For the measurements the *GetTickCount* function was used, which provides an accuracy of one millisecond.

[5] within the accuracy of measurement

The high value of Δt_{send} (max) is a result of the fact that the send thread is at the bottom of a dependency chain. *Hs_communicator* can only send data, after the Matlab calculations provide the result data and the calculations can only be performed, after the receive thread of *hs_communicator* provides the received data as input for the calculations.

The oscilloscope measurement was set up as follows: The *MaxLoadCE* thread was started with a certain priority, along with the RTBS. Different from the software measurement, the *MaxLoadCE* was not terminated after 10 seconds and ran until the external measurement was finished. The oscilloscope was connected to the output RS232 line of the real-time computer. Figure 4 shows pictures taken from the oscilloscope (width of one square equals 10ms).

The top picture in Fig. 4 shows the outgoing system data with no additional load. Outgoing data of the stressed system is shown in the middle picture. For this purpose *MaxLoadCE* was started with the highest possible priority below the real-time level. The picture at the bottom shows the outgoing system data, while *MaxLoadCE* is running at real-time priority. It is easy to see that a thread with a priority below real-time level is not able to affect the real-time processing in any way. Again the *MaxLoadCE* at real-time priority does disturb other real-time processes for 100 ms, but the data is sill processed, as shown in the bottom picture. Summarized Windows CE and the RTBS are showing the expected real-time behavior. Non real-time processing does not affect real-time processes. Beside the synthetic tests, the real-time behavior of the whole system was evaluated in flight experiments. The data recorded during these experiments shows no signs of temporal anomalies (e.g. unexpected level shifts of recorded sensor data).

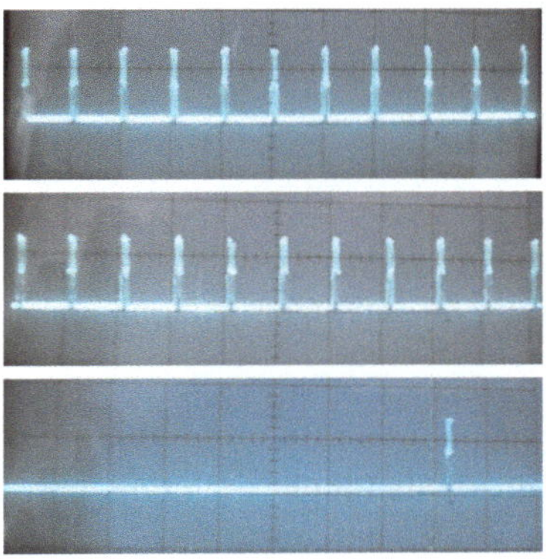

Fig. 4. Different result from oscilloscope measurement

6 Conclusions

In this paper, a framework for development and control of small size autonomous helicopters was presented. The main part of this framework is constituted by the embedded software called RTBS. RTBS can be considered as middleware composed of different processes and interface libraries which provide interface to the helicopter hardware as well as the communication and synchronization facilities. The communication facilities are implemented using the well known blackboard concept which was extended by synchronization events. The connection of the RTBS to the Matlab/Simulink makes the whole system suitable not only for operation of autonomous helicopters, but also for development and debugging of control software. The combination of Matlab/Simulink with RTBS allows usage of automatic code generation functionalities of Matlab with the helicopter hardware and with the low level software for actuators, sensors and communication. The real-time behavior of the system was shown in a special laboratory setup as well as in real flight experiments. The presented system has been used in the authors' lab for approximately two years. The usage of the presented system allowed the fast development and testing of different novel algorithms for autonomous helicopter control, s. e.g. [8].

References

1. C. Deeg, M. Musial and G. Hommel. "Control and Simulation of an Autonomously Flying Model Helicopter". 5th Symposium on Intelligent Autonomous Vehicles, 2004.
2. M. Bernard. "Entwurf und Implementierung einer Steuerung für kooperierende autonome Hubschrauber". Master thesis, TUB 2005.
3. M. Gergeleit and H. Streich. "Implementing a distributed high-resolution real-time clock using the can-bus", in 1st international CAN Conference, 1994
4. C. Liu, J. Layland. "Scheduling Algorithms for Multiprogramming in a Hard Real-Time Environment". Journal of the ACM 20, 1, 1973.
5. M. Thomson and J. Browne. "Designing and Optimizing Microsoft Windows CE .NET for Real-Time Performance". Microsoft Corp. study, 2002
6. M. Struys and M. Verhagen. "Real-Time Behaviour of the .NET Compact Framework", PTS Software study 2003
7. C. Tacke and L. Ricci. "Benchmarking Real-Time Determinism in Microsoft Windows CE". Applied Data Systems, 2002
8. K. Kondak, N. Losse, M. Bernard and G. Hommel. "Elaborated Modeling and Control for Autonomous Small Size Helicopters". ISR/Robotik 2006.

Embedded System Design for a Hand Exoskeleton

Andreas Wege and Günter Hommel

Real-Time Systems and Robotics
Technische Universität Berlin, Germany
{awege,hommel}@cs.tu-berlin.de

Summary. The focus of this paper is the description of the system design of a hand exoskeleton. This device is developed with focus on support of the rehabilitation process after hand injuries or strokes. As the device is designed for the later use on patients, which have limited hand mobility, fast undesired movements have to be averted. Safety precautions in the hardware and software design of the system must be taken to ensure this. An other important aspect is the integration of the control algorithms and data acquisition for the numerous degrees of freedom. The system currently can control 20 degrees of freedom through PWM motor amplifiers. For each degree of freedom there are multiple sensor inputs available. These have to be acquired and processed in real-time to generate the control output. The system design is based on a dedicated real-time controller running the real-time operation system Pharlap. The controller is equipped with four expansion cards supplying analogue input and output channels and two FPGA-boards. For programm development a combination of LabView and LabWindows/CVI was used. The controller connects to a host computer via Ethernet, which provides the user interface and sends the desired trajectory to the real-time controller.

1 Introduction and Motivation

After injuries of the hand or surgery, a rehabilitation process is necessary to regain as much dexterity back as possible. For example, rehabilitation can be required to prevent agglutination or adhesion of the involved tissue. After long periods of being unable to use the hand, it may also be required to relearn basic movements. This is also important for stroke patients. The rehabilitation process is time-consuming and labour-intensive, making the process expensive. If the rehabilitation is not performed optimal, the consequences can be serious. Especially limited hand dexterity can be a big problem for affected persons. Social and economic consequences can be severe.

Machines can support the process of rehabilitation. Currently only simple machines are available and in use. These apply a continuous motion [1].

Günter Hommel and Sheng Huanye (eds.), Embedded Systems Modeling – Technology, and Applications, 169–176.

The flexibility of these devices is limited; they support few degrees of freedom and provide no sensor data as feedback for the therapist.

Other devices used in virtual reality application look promising at first but lack of several features that are essential for rehabilitation. These devices often apply force only into one direction, whereas bidirectional force application is desirable for rehabilitation. Some only use brakes to restrict the motion. To simulate contacts in a virtual environment this is convenient and possibly even safer than moving the joint actively [2]. However, for patients incapable to move their hands it is inappropriate.

Some more sophisticated devices under development already have shown to be useful during rehabilitation [3, 4], although the number of actively powered joints is relatively low [5]. One the basis of these findings the development of the hand exoskeleton shown in Fig. 1 was started.

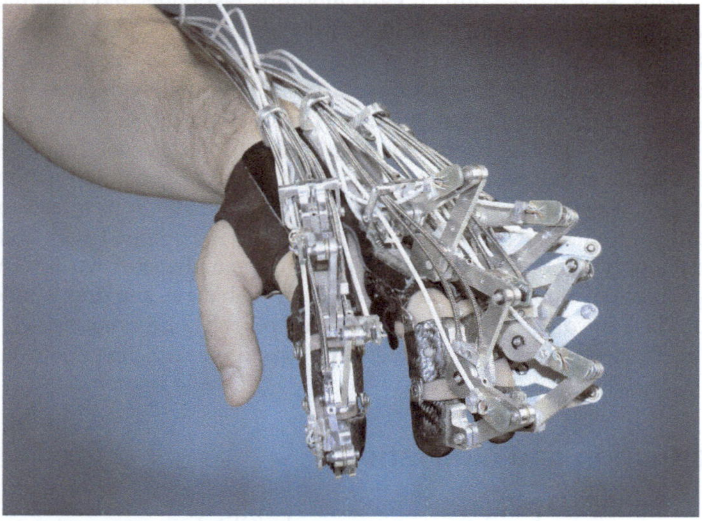

Fig. 1. Hand Exoskeleton with four fingers actuated in four degrees of freedom each.

2 System Requirements

To support rehabilitation the hand exoskeleton should be able to move the hands in multiple degrees of freedom. The system was designed to support up to 20 finger joints – four for each finger. Following system requirements have to be fulfilled:

- Support the data acquisition for sensors described in section 2.1
- Support control outputs for actuators described in section 2.2
- Supply enough calculation power to run control loops
- Safety consideration (avoid undesired movements)

The system design supporting these requirements is described in section 3. Section 4 draws conclusions and gives an outlook on possible extensions of the system.

2.1 Sensors of the System

The system is equipped with following sensors:

- Hall sensors to measure joint angles
- Optical quadrature encoders to measure angles of motor axes
- Force sensors measuring the force between human and construction
- Electromyography (EMG) sensors to measure muscle activity

To provide angular measurement the angle of the levers of the construction are measured by hall sensors (*KMZ41 from Philips*) which are evaluated by integrated circuits (*UZZ9001 from Philips*). These can be read-out through the serial peripheral interface (SPI). On the other side of the construction optical quadrature encoders (from Maxon) are used to measure the position of the motor axes. The values for the calculated joint angles deviate because of varying tension of the Bowden cables. Due to this redundant measurement, it is possible to detect mechanical failure by comparing the values. The joint angle measurement is implemented for all joints.

In addition to the joint angle measurement, force sensors (Entran ELFS-B3-50N) are integrated into the construction. The sensors are placed between the finger attachments and the levers and thereby measure the force transferred from the human to the construction. Due to mechanical reasons the angle of the force application to the sensor varies from about 70 to 40 degrees. Because sensors in use are calibrated for orthogonal force application, measured forces have to be corrected depending on the joint angles. Currently only one finger is equipped with force sensors but the system is designed for an extension on all fingers.

Electromyography (EMG) electrodes can measure the electric activity of up to eight muscles. The electrodes (*DE-2.3 MyoMonitor from DelSys*) have integrated signal amplifiers. As a matter of safety, EMG electrodes are powered by rechargeable batteries and must be galvanically insulated from the data acquisition circuit. Because of the high number of muscles in the forearm, recorded signals are superimposed. To improve the signal quality methods of blind source separation would have to be applied. Currently this is not done and the signals are only recorded for diagnostic purposes and not used for any control purposes.

2.2 Actuator Setup

The joints are driven trough Bowden cables by standard DC motors. Bidirectional movement is supported by the use of two pull cables for each joint diverted by a pulley on both ends. This allows using only one motor for each joint, but introduces some slackness compared to the solution using one motor for each direction. The motion is applied through a leverage construction onto each finger attachment. Details of the construction are described in [6].

Standard PWM controllers (*4122Z from Copley Controls*) drive the motors on the other side of the Bowden cables. These controllers are used in torque control mode and are driven by analogue inputs. Current monitor output allows observation of the PWM controller operation.

Each motor is equipped with limit switches on both maximum excitations. As soon as a limit switch is activated or a cable to the switch is disconnected the controller can stop the motion of the corresponding motor. By using an emergency switch, the power supply can also be interrupted at any time.

3 System Design Overview

The high number of sensors, controlled joints, real-time considerations, and safety consideration make it necessary to have a well-designed and reliable system. Direct utilisation of standard PC hardware with a Windows or Linux operation system was therefore not applicable.

For safety reasons the control loop and user interface are separated. The control loop runs on a dedicated real-time controller (sec. 3.1). The controller is supported by two FPGA-boards (sec. 3.2) and a watchdog constantly monitors the status of the controller (sec. 3.3). Through a communication protocol described in section 3.4 the control loop receives the desired trajectory from a separate standard PC running the user interface (sec. 3.5). Figure 2 gives an overview of the entire system.

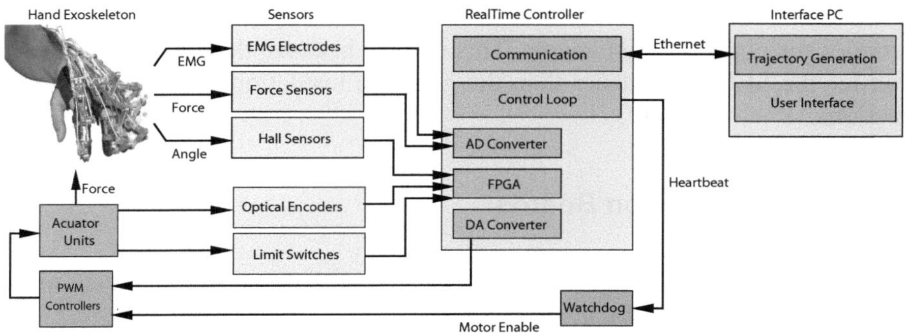

Fig. 2. System overview with major parts of the system.

3.1 Dedicated Real-Time Controller

The control loop runs at 1 KHz on the real-time controller and controls the 20 joints. High-gain PID control provide a stable low-level control loop. Details of the control algorithm are described in [6]. Sanity checks limit positions and speeds of the received trajectory from the host computer. This should ensure that no undesired movements are executed.

The PXI embedded controller (*NI PXI-8187 from National Instruments*) is powered by a Pentium-4M running at 2.4 GHz and is extended by following PXI cards:

- two NI PXI-6229: Multifunction Data-Acquisition (DAQ) boards providing each 16 differential analogue converters (ADC) with a resolution of 16 Bits, 4 analogue output converters (DAC) with 16 Bit resolution, and 48 Digital Input-/Output (DIO) channels
- one NI PXI-6289: High Accuracy DAQ board providing 16 differential ADCs with a resolution of 18 Bit, 4 DACs with 16 Bit resolution, and 48 DIO channels
- two NI PXI-7831R: FPGA boards equipped with 96 DIO channels, 8 ADCs with 16 Bit resolution, 8 DACs with 16 Bit resolution, and a one million gate FPGA, programmable through LabView or VHDL

The system is running on the real-time operation system Pharlap and can be programmed with LabView or by a C application programming interface (*CVI libraries*).

The DAQ boards provide 4 of 20 analogue inputs and outputs to the PWM controllers. The FPGA expansion boards provide the others. Because of small amplitudes of force sensors signals, the PXI-6289 card is used to sample these sensors, as this card has a higher accuracy.

An external ADC that connects through SPI bus to the FPGA board samples the eight EMG channels. This is done because the galvanic insulation for the SPI bus is easier than the galvanic insulation of analogue connections.

3.2 FPGA Expansion Boards

The FPGA boards are programmed to evaluate the 20 optical encoders and provide 21 SPI inputs for Hall and EMG sensors (in parallel). They also provide 16 of the 20 analogue outputs for the PWM controllers and capture the corresponding current monitor signals from these controllers. Additionally the digital outputs are used to individual activate the PWM controllers by a motor enable signal. The 40 limiting switches of the motors are read-out through the FPGA boards as well.

Currently the FPGA boards only provide SPI inputs, numerous counters, data inputs, and outputs. As a future extension it is possible to use them to provide a low-level control loop for the system. The advantage would be a reduced processing load on the RT-controller and possibly higher reliability. Disadvantage would be the necessity to use fixed-point arithmetic on the FPGA, which makes it necessary to modify the algorithms slightly. Also long compilation time slows down the process of implementing new control algorithms.

3.3 Watchdog for the Real-Time Controller

The PWM controllers are provided with control signals by the real-time controller. If they do not receive updated control signals in time, they continue their last control command. As a result undesired movements would occur. As a safety measure, an external watchdog monitors a heartbeat signal generated from the control loop in the real-time controller. As soon as the signal is interrupted for more than a predefined time span (currently 5 ms), the watchdog generates a reset signal, which is used to set the enable flip-flop to false. The flip-flop and each motor enable signal are combined by logical conjunction, so that all motors are disabled when it is set to false. The enable flip-flop has to be reset manually by the human operator. By that measure it is possible to handle timeout failures of the real-time controller.

3.4 Communication System

For the communication between the real-time controller and the host computer, a real-time communication system developed at the Technical University of Berlin is used [8]. The protocol provides secure and insecure slots to transmit data. Status data e.g. for the display of the current movement is transmitted via insecure slots. Important data like the desired trajectory that is transmitted from the host computer to the real-time controler uses secure slots.

Low-level communication uses the TCP protocol stack. The UDP protocol, which is more appropriate for real-time communication, could not be used because no interface is provided by the *CVI libraries from National Instruments* to access the UDP stack on the real-time controller. Furthermore, the TCP communication on the real-time controller cannot be used non-blocking. Therefore, a separate process with a lower priority than the high priority control loop performs the communication with the host computer. The control loop accesses the data through circular buffers. If the communication link is interrupted and no new trajectory data is received, the hand exoskeleton stops in the last received position.

3.5 Host Computer for User Interaction

The host computer provides the user interface that allows generating trajectories, provides visual feedback of the executed motion by displaying a 3D model of the current hand position. Because the user interface is separated from the control loop and sanity checks on the received trajectory are performed, it is possible to develop the software on the host computer with less consideration of safety issues.

4 Conclusion and Outlook

The presented system design supports all required features and still allows extensions by integrating more sensors, more degrees of freedom, or enhanced control algorithms. Safety precautions are implemented by redundant sensors, watchdog monitoring of the real-time controller, limit switches, and sanity checks on the desired trajectory. The separation of the security relevant control loop and the user interface makes development safer.

Next steps will be the integrating of other control modes into the system. In addition, the safety and reliability of the device has to be improved further to allow initial experiments with patients in the future. The transfer of additional functions from the real-time controller into the FPGA is also considered. The basic control loop implemented on the FPGA could improve the performance and safety of the system design further. The final step would be to abandon the real-time controller and to use the FPGA alone for the control purposes of the hand exoskeleton.

5 References

[1]. JACE H440 continuous passive motion device for fingers: http://www.jacesystems.com/products/hand_h440.htm

[2] Koyama, T., Yamano, I., Takermua, K., Maeno, T., "Development of an Ultrasonic Clutch for Multi-Fingered Exoskeleton using Passive Force Feedback for Dexterous Teleoperation", Proceedings of the 2003 IEEE/RSJ Int. Conference on Intelligent Robots and Systems, Las Vegas Nevada, 2003

[3] Avizzano, C.A., Marcheschi, S., Angerilli, M., Fotana, M., Bergamasco, M., A Multi-Finger Haptic Interface for Visually Impeared People, Proc. of IEEE Int. Workshop on Robot and Human Interactive Communication, Millbrae, California, USA, 2003

[4] Sarakoglou, I., Tsagarakis, N.G., Caldwell, D.G., Occupational and physical therapy Using a hand exoskeleton based exerciser, Proceedings of IEEE/RSJ Int. Conference on Intelligent Robots and Systems, 2004, Sendai, Japan

[5] Jack, D., Boian, R., Merians, A., Tremaine, M., Burdea, G., Adamovich, S., Recce, M., Poizner, M., Virtual Reality-Enhanced Stroke Rehabilitation In: IEEE Trans. on Neural Systems and Rehabilitation Engineering, Vol. 9, No. 3, Sept. 2001

[6] Wege, A., Hommel, G., Development and Control of a Hand Exoskeleton for Rehabilitation of Hand Injuries. Proc. of the 2005 IEEE/RSJ Int. Conf. On Intelligent Robots and Systems (IROS), pp. 3461-3466, ISBN 0-7803-8913-1, IEEE, Edmonton, Alberta, Canada, 2.-6.8.2005

[7] Kondak, K., Hommel, G., Design of a Robust High Gain Pid Motion Controller using Sliding Mode Theory. Proc. IEEE ICMA 2005 (Int. Conf. on Mechatronics and Automation), pp. 26-31, ISBN 0-7803-9045-9, IEEE, Niagara Falls, Ontario, Canada, 29.7.-1.8.2005

[8] Musial, M., Remuß, V., Hommel, G., Middleware for Distributed Embedded Real-Time Systems, Proc. of. Int. Workshop on Embedded Systems - Modeling, Technology and Applications, Springer, Berlin, 2006

Embedded Control System for a Powered Leg Exoskeleton

Christian Fleischer and Günter Hommel

Real-Time Systems and Robotics
Technische Universität Berlin, Germany
{fleischer,hommel}@cs.tu-berlin.de

Summary. In this paper an embedded control system for a powered orthosis is presented. The orthosis (as shown in Fig. 1) is used to support the thigh muscles during flexion and extension of the knee while performing common motions like getting up from a chair, walking, or climbing stairs. The user interface of the control system is implemented by evaluating EMG signals from significant thigh muscles to find out the intended motion of the subject. The intended motion is executed with a linear actuator to support the subject's own muscle force. Results from two experiments are presented. During the trials the torque support from the actuator illustrates the performance of the system.

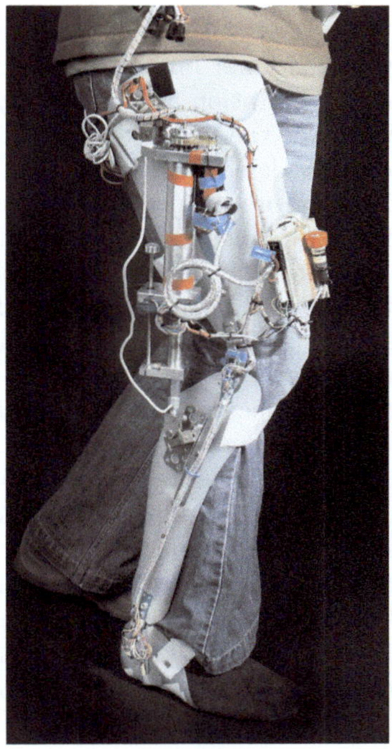

Fig. 1. The exoskeleton for knee support.

1 Introduction

Exoskeleton systems for human operators offer a wide range of possible applications: For patients they can offer assistance during their rehabilitation process by guiding motions on correct trajectories to help re-learning motion patterns, or give force support to be able to perform certain motions. In factory environments they could remove load from workers to avoid wearing down their bodies through strenuous physical work. Depending on the size,

Günter Hommel and Sheng Huanye (eds.), Embedded Systems Modeling – Technology, and Applications, 177–185.

weight, and handling of the devices, they could even be beneficial in everyday life at home, especially for elderly people.

Exoskeletons also offer a unique way of giving force feedback to the human body. They can act as haptic interfaces for telemanipulation, games and entertainment, or muscle and motion training devices for athletes.

Since numerous applications for exoskeletons can be imagined, many groups have shown interest in this topic. Especially in recent years several projects have emerged. The Berkley Lower Extremity Exoskeleton (BLEEX) for example is a military exoskeleton to aid soldiers carrying heavy loads [1]. The Hybrid Assistive Leg (HAL) is an actuated body suit for both legs [2], in the latest version (HAL-5) extended for both arms. It is designed for multiple purposes, such as supporting elderly people or as rehabilitation device. The powered lower limb orthosis described in [3] is developed to assist during motor rehabilitation after neurological injuries by re-learning gait patterns.

Whatever the use of such an exoskeleton is, there is need of an interface between the device and the person as soon as interaction is desired (no mere repetition of pre-defined trajectories).

Since standing, walking, and climbing stairs while keeping a stable pose are complex and very dynamic tasks, activating pre-defined motion patterns is not possible since perturbations from outside (contact forces) or the operator himself (swinging with the arms, leaning forward etc.) will interfere with the stability of the pre-calculated motions. So it is necessary to develop an intuitive interface that allows continuous control of the exoskeleton. In most of the developments one out of two alternatives is chosen: Either the desired motion is recognized by force sensors between the human and the mechanical construction or bio-signals are taken directly from the subject. The differences between the approaches are found in the underlying interpretation of the signals and the control algorithms.

Beside the standard application of EMG signals to analyze muscle activation and functionality during a rehabilitation process or analysis of diseases, more focus has recently been put on controlling robot arms and exoskeletons with those signals [4, 5, 6]. In Lloyd [7] a promising but very complex musculoskeletal model is presented that takes into account 13 muscles crossing the knee to estimate the resulting knee torque that could be used to control an exoskeleton, although the analysis presented there was not performed in real-time.

In [8] EMG signals of the Biceps Brachii and Triceps Brachii where used to estimate the elbow joint moment. With this estimation a moment controller was fed to allow control of a two-link exoskeletal arm to lift an external load with the hand. This approach is very similar to the one described here, but applied to a different environment.

The advantage of EMG signals compared to other means of input is that they form an intuitive interface and they can be used with every patient who is not paralyzed. Even if the muscles are not strong enough or the limbs hindered while performing a motion, signals of the intended motion can still be detected.

2 System Description

The control system can be divided into four parts: The sensor and actuator hardware, the microcontroller together with the control software, the PC for data visualization and user interaction, and the hardware safety system (refer to Fig. 2).

Sensor and Actuator System

The sensors integrated for reading the system state and user input are the knee angle sensor (Philips KMZ41 Hall sensor), the six EMG sensors (Delsys 2.3 differential electrodes), and the force sensor (GS Sensors XFTC300) in series with the linear actuator. This actuator consists of a ball screw powered by a standard dc motor (Maxon RE35, 90W) and is driven by a pwm amplifier (Copley 4122Z).

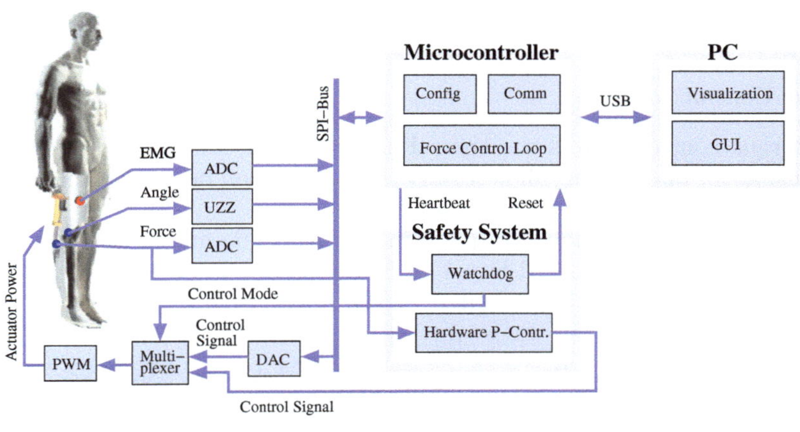

Fig. 2. Hardware of the control system. All sensor data is digitized and transmitted to the microcontroller via SPI bus. In parallel to the microcontroller a safety structure is implemented to control the orthosis in case of a microcontroller failure.

All those sensors and the actuator use converter circuits (MAX1230 ADC, AD5530 DAC, UZZ9001) to communicate with the microcontroller (Atmel Mega 32) via SPI bus. The microcontroller is responsible for taking the sensor information, computing the desired motion and sending appropriate control signals to the pwm amplifier. It also communicates with a PC by USB (non-real-time) to allow comfortable data visualization and user interaction.

The EMG signals are taken from the Rectus Femoris, Vastus Medialis, Vastus Lateralis, Semimembranosus, Semitendinosus, and Biceps Femoris. For the actual torque computation only three signals are used (Rectus Femoris, Vastus Medialis, Semimembranosus), the others are recorded to cross-check activation levels during development only.

The EMG signals are post-processed with offset elimination (average value of the last $200ms$ is taken as the offset), rectification and low-pass filtering. A second order Butterworth low-pass filter with a cutoff frequency between $2.0Hz$ to $4.0Hz$ (depending on the experiment and support ratio) was used. The post-processed EMG signal forms the activation envelope of the muscle j and is named $u_j(t)$.

Safety System

The safety system monitors the periodical heartbeat of the microcontroller. In case of a system fault the heartbeat stops and the separate hardware watchdog switches to a secondary controller: This controller is a simple hardware p-controller driven by the force sensor to avoid locking of the knee and allowing free motion (a force value of zero is the target). The EMG sensors have no influence here. The input of the pwm amplifier is switched from microcontroller to hardware p-controller with an analog multiplexer by the watchdog circuit. Switching back to software controlling has to be performed by hand for safety reasons.

Torque Control Loop

The torque control loop is the central aspect of the control structure. It is responsible for interpreting all necessary sensor data and producing the control value for the pwm amplifier (refer to Fig. 3).

To achieve this, the knee torque resulting from the muscle activations in the human thigh is estimated based on EMG signals. First, the force output for every single muscle j is calculated by the *EMG-to-Force function*:

$$F_{M,j}(u_j) = \frac{e^{A_j u_j u_{j,max}^{-1}} - 1}{e^{A_j} - 1} \cdot F_{j,max}, \tag{1}$$

where u_j is the activation of muscle j, $F_{j,max}$ is the force output when u_j is at its maximum ($u_j = u_{j,max}$), and A_j is a constant shape factor of the function with $-10 \leq A_j < 0$. A_j, $u_{j,max}$ and $F_{j,max}$ are calculated by a calibration routine as described in [9, 10].

The resulting knee torque is calculated in the *muscle model* by

$$t_{EMG}(u_j) = \sum_{j=1}^{N} \left((\mathbf{I}_j - \mathbf{J}) \times \frac{\mathbf{I}_j - \mathbf{O}_j}{|\mathbf{I}_j - \mathbf{O}_j|} \cdot F_{M,j}(u_j) \right), \tag{2}$$

where $\mathbf{J}$ is the vector to the knee joint, $\mathbf{O}_j$ and $\mathbf{I}_j$ the vectors to origin and insertion of the muscle in the reference frame. Obviously this sum allows co-contraction and co-activation but does not handle them explicitly.

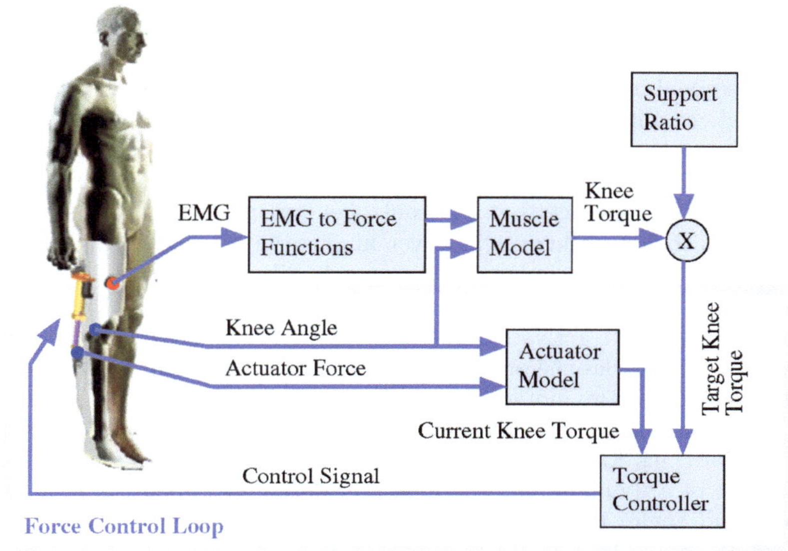

Fig. 3. Force control loop of the microcontroller.

This torque is multiplied by the *support ratio* r and interpreted as the *target torque* for the controller: $t_{target}(t) = t_{EMG}(t) \cdot r$. The *current knee torque* t_{cur} is derived from the force sensor that is connected between the actuator and the orthosis by a similar calculation as in Eq. 2. Both torques yield the formula for the control deviation of the *torque controller* (a standard p-controller to calculate the control signal for the pwm amplifier that drives the actuator): $E = t_{target} - t_{cur}$.

Properties of the Torque Control Loop

This simple control structure has some very interesting properties that will be described here:

For the whole system, only a few sensors are needed. Since human anatomy is very different from subject to subject, the less sensors are necessary the easier it is to adapt to different subjects and perform the calibration.

In contrast to algorithms where a physical model with dynamic equations is implemented, no knowledge about masses or velocities of the body parts is needed. All external and internal forces of the subject are summed up in the force sensor. No explicit measurement or calculation of those quantities is needed to perform the motion calculation. That results in a very robust system with no need for extra sensors for external reaction forces from e.g. the floor, a chair or a hand-rail (which are often difficulty to measure).

Another major difficulty is the unreliability of the EMG signals and the fact that deeper muscles cannot be recorded by surface electrodes. In case of a system with a body model that uses position controlled actuators with EMG sensors as the only input, the activation of muscles that are not recorded will not lead to any motion at all: the system will even prohibit the motion. The system described here will also not give any support when the muscles under observation are not activated, but it will not hinder the motion on the other hand, because the torque controller target value will be zero and the orthosis will be moved "out of the way" of the motion. That results in a very sensible behaviour: Given the fact that the controller can move the orthosis as fast as necessary, the exoskeleton will never hinder the subject during arbitrary motions (also if they are induced from outside, like pushing the leg with the hands) and it will support him or her with a specified ratio when muscles under observation are activated. Following that, depending on the activation of the different muscles the orthosis will only change the level of support but it will never hinder any motion.

A drawback of this kind of control loop is, that is is not possible to integrate algorithms for maintaining postural stability of the human. Due to the absence of a dynamic body model, no information about masses, accelerations, and angles is available, making it impossible to set up a system of dynamic equations that could take into account special constraints for the center of gravity and the zero moment point. It is also not possible to predict the motion (or even the direction of the motion) ahead of time to regard constraints like range of the knee angle etc.

3 Experiments

The experiments presented here have been performed with a healthy subject that is able to perform all motions on his own accord without any support. Due to safety issues it is currently not possible to perform experiments with patients with disabilities.

Nevertheless the results underline the usefulness of the system. The example motion presented here is climbing a stair. During the first experiment the support ratio was set to zero. The actuator was moving in accordance with the subject's leg, basically neither supporting nor hindering it. Analyzing the result of this setting is very useful to evaluate the controller parameters and the properties of the mechanical construction. Figure 4 shows the result from this experiment. The knee angle at the beginning is $-33deg$ ($0deg$ means straight leg, negative values indicate knee flexion). After that, the leg is raised to be put on the stair. At $t = 1.2s$ the leg is lowered onto the stair and the knee extensor muscles are getting activated. At $t = 1.7s$ the knee extensors are strongly activated to step up. At $t = 2.4s$ the knee is almost straight again and the activation of the extensor muscles ceases. The torque support of the actuator is almost zero during the whole motion. Except during the quick knee flexion in the interval of $0.7s < t < 1.2s$ where the actuator is a little too slow to keep up with the intended motion of the subject. A hindering torque with a maximum of $5Nm$ (positive, indicating knee flexion) is the result. This happens due to the limitation of the actuator current (for safety reasons) and the unfavourable geometric configuration regarding velocity at this point: The higher the knee flexion, the lower the ratio between linear velocity of the actuator and angular velocity of the knee joint (on the other hand the force transmission is higher in this configuration which is very desirable).

Figure 5 shows the same motion with a support ratio of 2.0. As can be seen, the curve of the knee angle is similar to the one without support, indicating that the system is still controllable by the user and the motion is still as desired. This is a very important fact, since the subject is also able to perform the task on his own. When adding a significant torque to the knee the human brain is fast enough to adapt to the different circumstances and is able to control the intended motion correctly. During push up between $2.0s < t < 2.6s$ the torque produced by the subject's muscles is reduced significantly and the torque contributed by the actuator is roughly two times larger than the estimated value from the human muscles as expected with a support ratio of 2.0.

Unfortunately it is not easy to compare the sum of the actuator torque and the muscle torque from the first trial with the second trial since different body poses and accelerations in the other joints affect the knee torque.

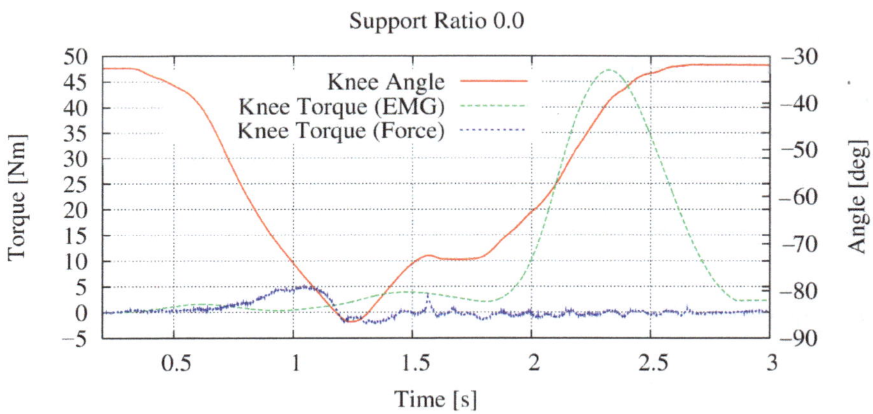

Fig. 4. Stair climbing experiment without support. The knee torque contribution from the actuator as measured by the force sensors is almost zero during the motion.

4 Results

The experiments have shown very good results for the stair climbing motion. Another positive example is getting up from a chair or sitting down. Those

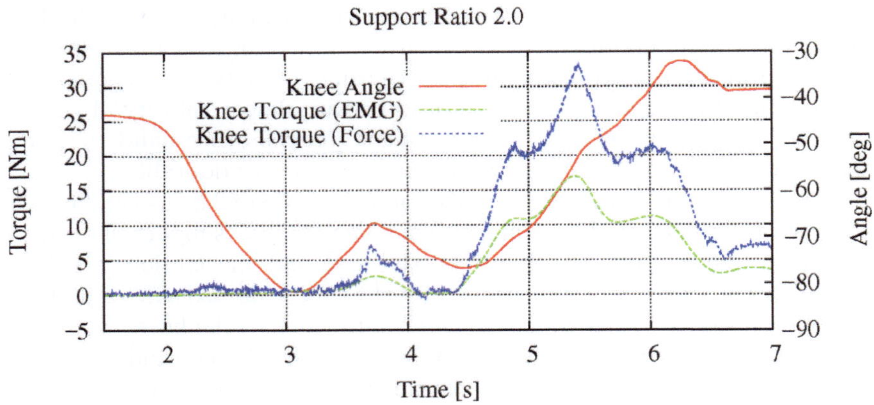

Fig. 5. Stair climbing experiment with 200% torque support, based on the estimated knee torque derived from the EMG signals.

motions (if executed normally) are not very dynamic and involve strong activation (and easy detection) of relevant muscle groups.

During walking the results are mixed: First of all, during normal walking thigh muscles are activated not very strong. The swinging of the shank is a result of the thigh motion created by the muscles responsible for hip flexion, extension and rotation. Thigh muscles are only used to "guide" the motion, to soften the reversal of motion and to stabilize the knee joint during floor contact. Those activations are very different from the ones during stair climbing and getting up and are harder to support. During floor contact the torque support works fine, but during the swing phase it rather leads to a motion that is not as smooth as desired. It might be sensible to turn off the force support during swing-phase. More work has to be performed in this area.

5 Conclusion

In this paper a system was presented to control an exoskeleton leg. Similar to the work in [8], the control loop utilizes a torque controller to add an adjustable support for the subject wearing the exoskeleton.

As shown with the experiments, using EMG signals as an input interface for a lower leg exoskeleton produces promising results for further investigations. They offer an intuitive way of commanding a control structure with the restriction that the user must be able to activate the muscles in a proper way to perform the desired motion.

The next steps of research will include making the motions more smooth and the EMG input safer by adding control layers to cope with undesired bursts. Once this and other safety features are implemented, experiments with patients can be performed.

References

1. H. Kazerooni, J.-L. Racinea, L. Huang, and R. Steger, "On the control of the berkley lower extremity exoskeleton (bleex)," in *Proceedings of the 2005 IEEE International Conference on Robotics and Automation*, 2005, pp. 4353–4360.
2. H. Kawamoto, S. Lee, S. Kanbe, and Y. Sankai, "Power assist method for hal-3 using emg-based feedback controller," in *Proceedings of the IEEE International Conference on Systems, Man and Cybernetics*, vol. 2, 2003, pp. 1648–1653.
3. G. S. Sawicki, K. E. Gordon, and D. P. Ferris, "Powered lower limb orthoses: Applications in motor adaption and rehabilitation," in *Proceedings of the IEEE International Conference on Rehabilitation Robotics*, vol. 2, 2005, pp. 206–211.
4. S. Lee and G. N. Sridis, "The control of a prosthetic arm by EMG pattern recognition," in *IEEE Transactions on Automatic Control*, vol. AC-29, no. 4, 1984, pp. 290–302.
5. O. Fukuda, T. Tsuji, H. Shigeyoshi, and M. Kaneko, "An EMG controlled human supporting robot using neural network," in *Proceedings of the IEEE/RSJ Int. Conf. on Intelligent Robots and Systems*, 1999, pp. 1586–1591.
6. S. Morita, T. Kondo, and K. Ito, "Estimation of forearm movement from EMG signal and application to prosthetic hand control," in *Proceedings of the IEEE Int. Conf. on Robotics & Automation*, 2001, pp. 3692–3697.
7. D. G. Lloyd and T. F. Besier, "An emg-driven musculoskeletal model to estimate muscle forces and knee joint moments in vivo," *Journal of Biomechanics*, vol. 36, pp. 765–776, 2003.
8. J. Rosen, M. Brand, M. B. Fuchs, and M. Arcan, "A myosignal-based powered exoskeleton system," in *IEEE Transactions on Systems, Man, and Cybernetics*, vol. 31, 2001.
9. C. Fleischer, K. Kondak, C. Reinicke, and G. Hommel, "Online calibration of the emg-to-force relationship," in *Proceedings of the IEEE/RSJ Int. Conf. on Intelligent Robots and Systems*, 2004.
10. C. Fleischer and G. Hommel, "Predicting the intended motion with emg signals for an exoskeleton orthosis controller," in *Proceedings of the IEEE/RSJ Int. Conf. on Intelligent Robots and Systems*, 2005.

Blind Source Separation of Temporal Correlated Signals and its FPGA Implementation

Xia Bin, and Zhang Liqing*

Perception Computing
SJTU-TUB Joint Institute for Information & Communication Technologies
Shanghai Jiao Tong University, China
{xbin,lqzhang}@sjtu.edu.cn

Summary. In this paper, we present a new framework for blind source separation (BSS). The difference between proposed method and original blind source separation method is that the source separation is performed in the residual level. We discuss two types of BSS problem: instantaneous BSS and convolutive BSS. The cost function is derived by simplifying the mutual information of residual signals for both problems. And then we develop efficient learning algorithms respectively. FPGA implementation of the proposed algorithm based on Signal-Master platform is also discussed. Computer simulations are given to show the separation performance of the proposed algorithm and some comparisons with other algorithms are also provided.

1 Introduction

Recently, natural human computer interactions (HCI) demand highly reliable multimodal interface technology. The speech recognition is one of important technologies in HCI. A number of commercial speech recognition systems with various vocabulary size are currently available. However the performance of those systems usually degrades substantially under real-world conditions. To improve the performance, we should enhance speech signals from recorded mixtures. Independent Component Analysis (ICA) or Blind Source Separation (BSS) is considered as an promising enhancement approach. The main idea of ICA is to retrieve the source signals from measure signals by making the output signals mutually independent. Hérault and Jutten [8] first introduced the concept of independent component analysis. In recent years, a number of ICA algorithms are developed [5, 7, 9, 11]. However, the high order statistical methods usually provide an over-separating solution if there are not enough training samples or there is a considerable amount of noise present [1, 10].

To overcome the drawback of over-separation, the high order statistics and temporal structures of source signals are both used to recover source signals [6, 3]. The

* The work was supported by the National Basic Research Program of China (Grant No. 2005CB724301) and National Natural Science Foundation of China (Grant No.60375015).

Günter Hommel and Sheng Huanye (eds.), Embedded Systems Modeling – Technology, and Applications, 187–195.

purpose of using the temporal structure is to deal with the over-separation problem and to improve the separating performance. In this paper, we propose a theoretical framework of blind source separation by separating sources in the residual level. We discuss the framework in instantaneous case at first. And then , we extend it to convolutive BSS. For real time application, we will briefly discuss field programmable gate array (FPGA) implementation of proposed BSS algorithm. Finally, computer simulations are given to demonstrate the separating performance of the proposed algorithm.

2 Problem Formulation

In this section, we describe the source signals, which has the temporally correlated structures, by using AR model. The mixing model is formulated in instantaneous BSS and convolutive BSS respectively. The objective functions for both case are derived from the mutual information of the residual signals.

2.1 Source Model

We assume that $s_i, i = 1, \ldots, n$ are n statistically independent information source signals and $s_i(k)$ are correlated at different time k. We describe the source signals **s** as follow

$$s_i(t) = \sum_{p=1}^{L} a_p^i s(k-p) + \varepsilon_i(k) \tag{1}$$

where L is the length of the FIR filter and $\varepsilon_i(t)$ is a zero mean independently and identically distributed time series called the residual. By introducing a delay operator z^{-1}, we can rewrite the Eq. (1) as

$$A_i(z)s_i(k) = \varepsilon_i(k) \tag{2}$$

where

$$A_i(z) = 1 - \sum_{p=1}^{L} a_p^i z^{-p} \tag{3}$$

In ordinary ICA model, only the demixing matrix parameters would be estimated. But in this paper, we describe the temporal correlated signals by introducing a minimum phase filter $A(z)$. So we should not only estimate the demixing matrix, but also obtain the filter $A(z)$.

2.2 Mixing Model of Instantaneous BSS

We use the generative model to describe the BSS problem. For the instantaneous case, a matrix can be use to represent the mixing model. For the simplicity, we assume the number of senors is equal to number of source signals. We present a two input and two output sample as follow

$$\begin{bmatrix} x_1(k) \\ x_2(k) \end{bmatrix} = \begin{bmatrix} h_{11} \; h_{12} \\ h_{21} \; h_{22} \end{bmatrix} \begin{bmatrix} s_1(k) \\ s_2(k) \end{bmatrix} + \begin{bmatrix} \xi_1(k) \\ \xi_2(k) \end{bmatrix} \tag{4}$$

where $x_i(t)$ is the sensor signal and $s_i(t)$ is the source signal. h_{ii} is scalar which is the entry of mixing matrix. ξ_i is white noise. For further describing the mixing model, we express Eq. (4) in a general form

$$\mathbf{x}(k) = \mathbf{Hs}(k) + \boldsymbol{\xi}(k) \tag{5}$$

where $\mathbf{x}(k)$ is vector of the mixing signals and $\mathbf{H}$ is the instantaneous mixing matrix. $\mathbf{s}(k)$ is vector of the mixing signals. $\boldsymbol{\xi} = (\xi_i, \ldots, \xi_n)^T$ is the vector of noises.

Blind source separation is to find a demixing matrix $\mathbf{W}$ which transforms observed $\mathbf{x}(k)$ to

$$\mathbf{y}(k) = \mathbf{Wx}(k) \tag{6}$$

where $\mathbf{y}(k)$ is the estimate of the source signals $\mathbf{s}(k)$. The $\mathbf{W}$ is called a separating solution, if it satisfies the follow condition

$$\mathbf{WA} = \boldsymbol{\Lambda}\mathbf{P}. \tag{7}$$

where $\boldsymbol{\Lambda} \in \mathbf{R}^{n \times n}$ is a nonsingular diagonal matrix and $\mathbf{P} \in \mathbf{R}^{n \times n}$ is a permutation

2.3 Mixing Model of Convolutive BSS

For the convolutive BSS case, we can use same form of Eq. (4) to describe the mixing model.

$$\begin{bmatrix} x_1(k) \\ x_2(k) \end{bmatrix} = \begin{bmatrix} \mathbf{h}_{11} \; \mathbf{h}_{12} \\ \mathbf{h}_{21} \; \mathbf{h}_{22} \end{bmatrix} * \begin{bmatrix} s_1(k) \\ s_2(k) \end{bmatrix} + \begin{bmatrix} \xi_1(k) \\ \xi_2(k) \end{bmatrix} \tag{8}$$

where $\mathbf{h}_{ii}$ is a filter and $*$ is a convolution operator. If we assume the mixing system is a minimum phase system, the mixing model can be describe as matrix form

$$\mathbf{x}(k) = \sum_{p=0}^{\infty} \mathbf{H}_p \mathbf{s}(k-p) + \boldsymbol{\xi}(k), \tag{9}$$

where $\mathbf{H}$ is an $n \times n$-dimensional matrix of mixing coefficients at time-lag p, which is called the impulse response at time p, $\mathbf{s}(k) = [s_1(k), \ldots, s_n(k)]^T$ is an n-dimensional vector of source signals with mutually independent components and $\mathbf{x}(k) = [x_1(k), \ldots, x_n(k)]^T$ is the vector of the sensor signals. Substituting the delay operator z into (9), we get the compact form of mixing model

$$\mathbf{x}(k) = \mathbf{H}(z)\mathbf{s}(k) + \boldsymbol{\xi}(k), \tag{10}$$

In many practical applications, source signals spread through multi-path. It will cause that observed signals generated by mixing together several delayed versions of the source signals. Therefore using the deconvolution model to describe mixing model is reasonable.

In blind deconvlution, the source signals $\mathbf{s}(k)$ and mixing coefficients of $\mathbf{H}(z)$ are unknown. The problem is to recovery the source signals $\mathbf{s}(k)$ or identify the channel $\mathbf{H}(z)$ only using observed signals $\mathbf{x}(k)$ and some statistical feature of source signals. The basic solution is to find deconvolution filter

$$\mathbf{y}(k) = \sum_{p=0}^{\infty} \mathbf{W}_p \mathbf{x}(k-p), \tag{11}$$

where $\mathbf{y}(k) = [y_1(k), \ldots, y_n(k)]^T$ is an n-dimensional vector of the outputs and $\mathbf{W}_p$ is an $n \times n$-dimensional coefficient matrix at time-lag p. Similarly, the compact form of demixing model can be described as $\mathbf{y}(k) = \mathbf{W}(z)\mathbf{x}(k)$.

In independent component analysis (ICA), there exist scaling ambiguity and permutation ambiguity [2] because some prior knowledge of source signals are unknown. Similarly, these indeterminacies remain in the blind deconvolution problem. We can rewrite (11) as

$$\mathbf{y}(k) = \mathbf{W}(z)\mathbf{x}(k) = \mathbf{W}(z)\mathbf{H}(z)\mathbf{s}(k) = \mathbf{P}\mathbf{\Lambda}\mathbf{D}(z)\mathbf{s}(k), \tag{12}$$

where $\mathbf{P} \in \mathbf{R}^{n \times n}$ is a permutation matrix, $\Lambda \in \mathbf{R}^{n \times n}$ is a nonsingular diagonal scaling matrix. Then the global transfer function is defined by $\mathbf{G}(z) = \mathbf{W}(z)\mathbf{H}(z)$. The blind deconvolution task is to find a demixing filter $\mathbf{W}(z)$ which satisfies following condition

$$\mathbf{G}(z) = \mathbf{W}(z)\mathbf{H}(z) = \mathbf{P}\mathbf{\Lambda}\mathbf{D}(z), \tag{13}$$

where $\mathbf{D}(z) = diag\{z^{-d_1}, \ldots, z^{-d_n}\}$.

3 Cost Function

In this section, we introduce the mutual information of residual signals as a criterion for training the demixing matrix and temporal structure parameters. The main purpose of this paper is to train the demixing matrix and the temporal filter so as to minimize the mutual information of the residual signals.

From source model Eq. (2), we have

$$\varepsilon(k) = \mathbf{A}(z)\mathbf{s}(k) \tag{14}$$

where $\mathbf{A}(z) = diag(A_q(z), \ldots, A_n(z))^T$ can be estimated via the linear prediction method if the source signals $\mathbf{s}(k)$ are know. However, in the blind separation problem, we neither know the source signal nor their temporal structures. If we can estimate the temporal structure $\mathbf{A}(z)$ and demixing model well, the residual signals of output can be described as follow

$$\mathbf{r}(k) = \mathbf{A}(z)\mathbf{W}x(k), \quad instantaneous \quad case \tag{15}$$

$$\mathbf{r}(k) = \mathbf{A}(z)\mathbf{W}(z)x(k), \quad convolution \quad case \tag{16}$$

The $\mathbf{r}(k)$ is mutually independent. In ordinary ICA model, the mutual information is choose as train criterion for making the output signals mutually independent. Here we do not make output signals mutually independent ,but make residual output signals mutually independent.

If the random variables $r_1, \ldots, r_n$ are mutually independent, we have

$$q(\mathbf{r}) = \prod_{i=1}^{n} q_i(r_i). \tag{17}$$

where $q(\mathbf{r})$ is the probability density function of $\mathbf{r}$ and $q_i(r_i)$ is the marginal probability density function of $r_i, i = 1, \ldots, n.$

In [4], Amari *et al.* given a simple cost function as follows

$$L(\mathbf{y}, \mathbf{A}(z), \mathbf{W}(z)) = -\frac{1}{2\pi j}\oint_{\gamma} \log|det(\mathbf{W}(z))|z^{-1}dz - \sum_{i=1}^{n} E[\log p_i(r_i)], \tag{18}$$

In this paper, the first term of above equation is

$$\log|det(\mathbf{A}(z)\mathbf{W}(z))| = \log|det(\mathbf{A}(z))| + \log|det(\mathbf{W}(z))| \tag{19}$$

In our previous assumption, the temporal filter $\mathbf{A}(z)$ is causal and minimum phase, we can easily verify

$$\frac{1}{2\pi j}\oint_{\gamma} \log|det(\mathbf{A}(z))|z^{-1}dz = 0 \tag{20}$$

As we assume the demixing filter $\mathbf{W}(z)$ is minimum phase before, so we can the second term of right side of (19) as follows:

$$\frac{1}{2\pi j}\oint_{\gamma} \log|det(\mathbf{W}(z))|z^{-1}dz = \log|det(\mathbf{W}_0)|, \tag{21}$$

So the cost function (18) can be rewrote as follows

$$L(\mathbf{y}, \mathbf{A}(z), \mathbf{W}(z)) = -\log|det(\mathbf{W}_0)| - \sum_{i=1}^{n} \log p_i(r_i) \tag{22}$$

For instantaneous case, we obtain the similar cost function

$$L(\mathbf{y}, \mathbf{A}(z), \mathbf{W}) = -\log|det(\mathbf{W})| - \sum_{i=1}^{n} \log p_i(r_i) \tag{23}$$

It should be noted here that the activation functions are functions of the residual variables, rather than the source variables.

4 Learning Algorithm

In this section, we develop the algorithm for both instantaneous and convolutive cases. We only discuss the instantaneous case here. In the same way, we can develop the algorithm for convolutive case.

The differential form of the cost function (23) is expressed as:

$$dl(\mathbf{y}, \mathbf{A}, \mathbf{W}(z)) = -d\log|det(\mathbf{W})| - \sum_{i=1}^{n} d\log p_i(r_i) \tag{24}$$

The first term of above equation can be described as

$$d\log(|det(\mathbf{W})|) = tr(d\mathbf{W}\mathbf{W}^{-1}) \tag{25}$$

The second term can be rewrote as

$$\begin{aligned} -\sum_{i=1}^{n} d\log p_i(r_i) &= \varphi(\mathbf{r})^T d\mathbf{r} \\ &= \varphi(\mathbf{r})^T \mathbf{A}(z) d\mathbf{W}\mathbf{W}^{-1}\mathbf{W}\mathbf{x} \\ &= \varphi(\mathbf{r})^T \mathbf{A}(z) d\mathbf{W}\mathbf{W}^{-1}\mathbf{y} \end{aligned} \tag{26}$$

We introduce an non-holonomic transform for the matrix $\mathbf{W}$

$$d\mathbf{X} = d\mathbf{W}\mathbf{W}^{-1} \tag{27}$$

By substituting (25),(26) and (27) in (24), we obtain

$$\begin{aligned} dl(\mathbf{y}, \mathbf{A}(z), \mathbf{W}) &= -tr(d\mathbf{X}) + \varphi(\mathbf{r})^T \mathbf{A}(z) d\mathbf{X}\mathbf{y}(k) \\ &= -tr(d\mathbf{X}) + \varphi(\mathbf{r})^T \sum_{p=0}^{L} \mathbf{A}_p d\mathbf{X}\mathbf{y}(k) \end{aligned} \tag{28}$$

We can calculate the derivatives of the cost function with respect to $\mathbf{X}$

$$\frac{\partial l(\mathbf{y}, \mathbf{A}(z), \mathbf{W})}{\partial \mathbf{X}} = -\mathbf{I} + \sum_{p=0}^{L} \mathbf{A}_p^T \varphi(\mathbf{r}) \mathbf{y}^T(k-p) \tag{29}$$

Combining (27) and (29), the update rule of demixing filter $\mathbf{W}$ is obtained as

$$\triangle \mathbf{W} = \eta \left\{ \mathbf{I} - \sum_{p=0}^{L} \left[\mathbf{A}_p \varphi(\mathbf{r}) \mathbf{y}^T(k-p) \right] \right\} \mathbf{W}. \tag{30}$$

Similarly, we can obtain the algorithm of convolutive case as:

$$\begin{aligned} \triangle \mathbf{W}_p &= -\eta \sum_{q=0}^{p} \frac{\partial l(\mathbf{y}, \mathbf{A}(z), \mathbf{W}(z))}{\partial \mathbf{X}_q} \mathbf{W}_{p-q} \\ &= -\eta \sum_{q=0}^{p} \left((-\delta_{0,q}\mathbf{I} + \sum_{i=0}^{L} \mathbf{A}_i^T \varphi(\mathbf{r}) \mathbf{y}^T(k-i-q)) \right) \mathbf{W}_{p-q} \end{aligned} \tag{31}$$

From Eq. (30), (31), we know there exist an unknown temporal filter A(z). In this paper, we use linear predictor method to determine the coefficients. Due to limit of space, we do not discuss it here. In one full paper of future, we will present the details of how to derive the update rules of $\mathbf{A}(z)$ and how to develop the algorithm (31).

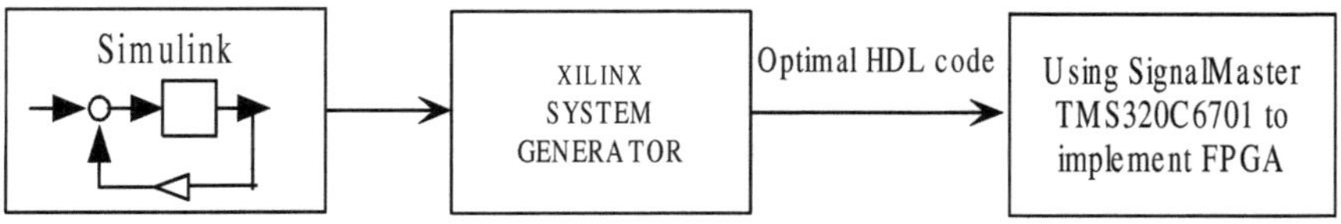

Fig. 1. Illustration of FPGA implementation flow

5 Discussion of FPGA Implementation

In previous section, we developed algorithms for blind source separation problem. As we known,algorithm (30) and (31) are both iterative algorithms. They will take long time to achieve convergence. Especially in convolutive BSS case, there are several hundred taps in demixing filter $\mathbf{W}(z)$ due to multipath reverberative environment. In real time application, we need more powerful computing ability to obtain fast convergence and good separation performance. FPGA architecture provides high computing power and offers rapid hardware prototyping and algorithm verification. In this project, we plan to use SignalMaster as DSP/FPGA development platform. Figure1 shows the implementation flow. We perform simulation of the proposed algorithms in simulink environment first. Then generating optimal HDL code by using xilinx system generator tool. Finally, we input HDL code in SignalMaster to implement the BSS algorithms for real time application. The details of FPGA design will be discussed in another paper.

6 Simulation

We now present simulation examples to illustrate the performance of the proposed algorithms. Simulation one is to illustrate the separation performance of proposed algorithm for speech signals. Figure 2 shows separation performance of proposedalgorithm. Top row is source signals. We use a random mixing matrix **A** to mix them together, as show in middle row. Using proposed algorithm to demix the source, bottom row shows we obtain good separation performance.

To compare the separation performance other methods, we also apply three other algorithms: SOBI-RO, SONS and JADE to separate the speech signals. And making them work at different SNR condition. Figure 3 shows the proposed algorithm achieve the best separation result.

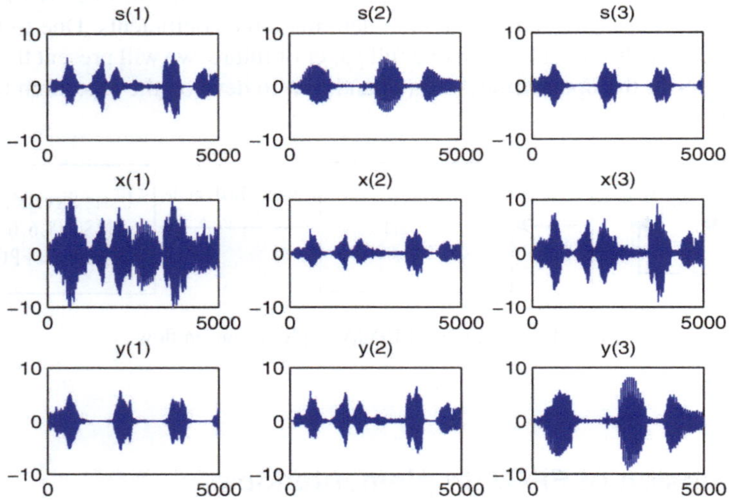

Fig. 2. Illustration of separation performance

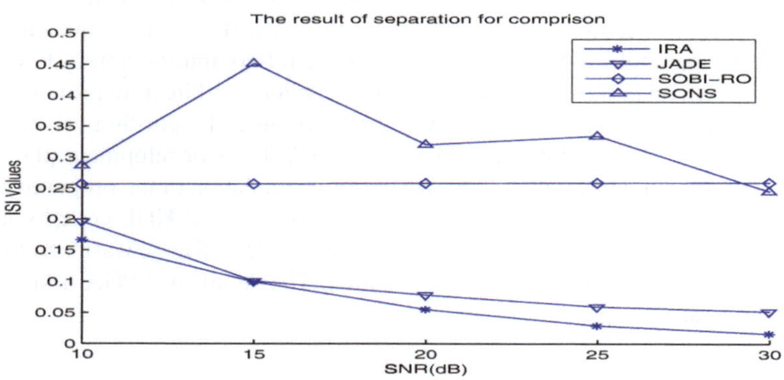

Fig. 3. Illustration of comparison result

7 Conclusions

In this paper, we present a new framework of blind source separation. We discuss the both instantaneous and convolutive case. The new framework suggests to use both the high-order statistics and temporal structures of the source signals in separation algorithm. The main idea of the proposed framework is to separate source signals in the residual level. We develop effective algorithms for both cases. Due to limit of space, we only present some simulation in instantaneous case. More discussion of convolutive BSS case and FPGA implementation will be presented in a full paper.

Computer simulations show that the proposed instantaneous BSS algorithm works well.

References

1. J. Sarela, A. Hyvarinen, and R. Vigario. Spikes and bumps: Artefacts generated by independent component analysis with insufficient sample size. In *Proc. Int Workshop on Independent Component Analysis and Blind Signal Separation, ICA'99*, pages 425–429, Aussois, France, 1999.
2. A. Hyvärinen, J. Karhunen, and E. Oja. *Independent component analysis*. Wiley, 2001.
3. S. Amari. Estimating function of independent component analysis for temporally correlated signals. *Neural Computation*, 12(9):2083–2107, September 2000.
4. S. Amari and A. Cichocki. Adaptive blind signal processing - neural network approaches. *Proceedings of the IEEE*, 86(10):2026–2048, 1998.
5. S. Amari, A. Cichocki, and H. H. Yang. A new learning algorithm for blind signal separation. In G. Tesauro, D. S. Touretzky, and T. K. Leen, editors, *Advances in Neural Information Processing Systems 8(NIPS 95)*, pages 757–763, Cambridge, MA, 1996. The MIT Press.
6. H. Attias and C. Schreiner. Blind source separation and deconvolution: The dynamic component analysis algorithm. *Neural Computation*, 10(6):1373–1424, 1998.
7. P. Comon. Independent component analysis: a new concept? *Signal Processing*, 36: 287–314, 1994.
8. J. Hérault and C. Jutten. Space or time adaptive signal processing by neural network models. In *Neural networks for computing: Proceedings of the AIP Conference*, pages 206–211, New York, 1986. American Institute of Physics.
9. A. Hyvärinen and E. Oja. A fast fixed-point algorithm for independent component analysis. *Neural Computation*, 9:1483–1492, 1997.
10. J. Sarela and R. Vigario. Overlearning in marginal distribution-based ica: Analysis and solutions. *Journal of Machine Learning Research*, 4:1447–1469.
11. L. Q. Zhang, A. Cichocki, and S. Amari. Natural gradient algorithm for blind separation of overdetermined mixture with additive noise. *IEEE Signal Processing Letters*, 8(11):293–295, 1999.

Mechanics

SOLID MECHANICS AND ITS APPLICATIONS

Series Editor: G.M.L. Gladwell

Aims and Scope of the Series

The fundamental questions arising in mechanics are: *Why?*, *How?*, and *How much?* The aim of this series is to provide lucid accounts written by authoritative researchers giving vision and insight in answering these questions on the subject of mechanics as it relates to solids. The scope of the series covers the entire spectrum of solid mechanics. Thus it includes the foundation of mechanics; variational formulations; computational mechanics; statics, kinematics and dynamics of rigid and elastic bodies; vibrations of solids and structures; dynamical systems and chaos; the theories of elasticity, plasticity and viscoelasticity; composite materials; rods, beams, shells and membranes; structural control and stability; soils, rocks and geomechanics; fracture; tribology; experimental mechanics; biomechanics and machine design.

1. R.T. Haftka, Z. Gürdal and M.P. Kamat: *Elements of Structural Optimization*. 2nd rev.ed., 1990 ISBN 0-7923-0608-2
2. J.J. Kalker: *Three-Dimensional Elastic Bodies in Rolling Contact*. 1990 ISBN 0-7923-0712-7
3. P. Karasudhi: *Foundations of Solid Mechanics*. 1991 ISBN 0-7923-0772-0
4. *Not published*
5. *Not published.*
6. J.F. Doyle: *Static and Dynamic Analysis of Structures*. With an Emphasis on Mechanics and Computer Matrix Methods. 1991 ISBN 0-7923-1124-8; Pb 0-7923-1208-2
7. O.O. Ochoa and J.N. Reddy: *Finite Element Analysis of Composite Laminates*. ISBN 0-7923-1125-6
8. M.H. Aliabadi and D.P. Rooke: *Numerical Fracture Mechanics*. ISBN 0-7923-1175-2
9. J. Angeles and C.S. López-Cajún: *Optimization of Cam Mechanisms*. 1991 ISBN 0-7923-1355-0
10. D.E. Grierson, A. Franchi and P. Riva (eds.): *Progress in Structural Engineering*. 1991 ISBN 0-7923-1396-8
11. R.T. Haftka and Z. Gürdal: *Elements of Structural Optimization*. 3rd rev. and exp. ed. 1992 ISBN 0-7923-1504-9; Pb 0-7923-1505-7
12. J.R. Barber: *Elasticity*. 1992 ISBN 0-7923-1609-6; Pb 0-7923-1610-X
13. H.S. Tzou and G.L. Anderson (eds.): *Intelligent Structural Systems*. 1992 ISBN 0-7923-1920-6
14. E.E. Gdoutos: *Fracture Mechanics*. An Introduction. 1993 ISBN 0-7923-1932-X
15. J.P. Ward: *Solid Mechanics*. An Introduction. 1992 ISBN 0-7923-1949-4
16. M. Farshad: *Design and Analysis of Shell Structures*. 1992 ISBN 0-7923-1950-8
17. H.S. Tzou and T. Fukuda (eds.): *Precision Sensors, Actuators and Systems*. 1992 ISBN 0-7923-2015-8
18. J.R. Vinson: *The Behavior of Shells Composed of Isotropic and Composite Materials*. 1993 ISBN 0-7923-2113-8
19. H.S. Tzou: *Piezoelectric Shells*. Distributed Sensing and Control of Continua. 1993 ISBN 0-7923-2186-3
20. W. Schiehlen (ed.): *Advanced Multibody System Dynamics*. Simulation and Software Tools. 1993 ISBN 0-7923-2192-8
21. C.-W. Lee: *Vibration Analysis of Rotors*. 1993 ISBN 0-7923-2300-9
22. D.R. Smith: *An Introduction to Continuum Mechanics*. 1993 ISBN 0-7923-2454-4
23. G.M.L. Gladwell: *Inverse Problems in Scattering*. An Introduction. 1993 ISBN 0-7923-2478-1

Mechanics

SOLID MECHANICS AND ITS APPLICATIONS

Series Editor: G.M.L. Gladwell

24. G. Prathap: *The Finite Element Method in Structural Mechanics*. 1993 ISBN 0-7923-2492-7
25. J. Herskovits (ed.): *Advances in Structural Optimization*. 1995 ISBN 0-7923-2510-9
26. M.A. González-Palacios and J. Angeles: *Cam Synthesis*. 1993 ISBN 0-7923-2536-2
27. W.S. Hall: *The Boundary Element Method*. 1993 ISBN 0-7923-2580-X
28. J. Angeles, G. Hommel and P. Kovács (eds.): *Computational Kinematics*. 1993 ISBN 0-7923-2585-0
29. A. Curnier: *Computational Methods in Solid Mechanics*. 1994 ISBN 0-7923-2761-6
30. D.A. Hills and D. Nowell: *Mechanics of Fretting Fatigue*. 1994 ISBN 0-7923-2866-3
31. B. Tabarrok and F.P.J. Rimrott: *Variational Methods and Complementary Formulations in Dynamics*. 1994 ISBN 0-7923-2923-6
32. E.H. Dowell (ed.), E.F. Crawley, H.C. Curtiss Jr., D.A. Peters, R. H. Scanlan and F. Sisto: *A Modern Course in Aeroelasticity*. Third Revised and Enlarged Edition. 1995 ISBN 0-7923-2788-8; Pb: 0-7923-2789-6
33. A. Preumont: *Random Vibration and Spectral Analysis*. 1994 ISBN 0-7923-3036-6
34. J.N. Reddy (ed.): *Mechanics of Composite Materials*. Selected works of Nicholas J. Pagano. 1994 ISBN 0-7923-3041-2
35. A.P.S. Selvadurai (ed.): *Mechanics of Poroelastic Media*. 1996 ISBN 0-7923-3329-2
36. Z. Mróz, D. Weichert, S. Dorosz (eds.): *Inelastic Behaviour of Structures under Variable Loads*. 1995 ISBN 0-7923-3397-7
37. R. Pyrz (ed.): *IUTAM Symposium on Microstructure-Property Interactions in Composite Materials*. Proceedings of the IUTAM Symposium held in Aalborg, Denmark. 1995 ISBN 0-7923-3427-2
38. M.I. Friswell and J.E. Mottershead: *Finite Element Model Updating in Structural Dynamics*. 1995 ISBN 0-7923-3431-0
39. D.F. Parker and A.H. England (eds.): *IUTAM Symposium on Anisotropy, Inhomogeneity and Nonlinearity in Solid Mechanics*. Proceedings of the IUTAM Symposium held in Nottingham, U.K. 1995 ISBN 0-7923-3594-5
40. J.-P. Merlet and B. Ravani (eds.): *Computational Kinematics '95*. 1995 ISBN 0-7923-3673-9
41. L.P. Lebedev, I.I. Vorovich and G.M.L. Gladwell: *Functional Analysis*. Applications in Mechanics and Inverse Problems. 1996 ISBN 0-7923-3849-9
42. J. Menčik: *Mechanics of Components with Treated or Coated Surfaces*. 1996 ISBN 0-7923-3700-X
43. D. Bestle and W. Schiehlen (eds.): *IUTAM Symposium on Optimization of Mechanical Systems*. Proceedings of the IUTAM Symposium held in Stuttgart, Germany. 1996 ISBN 0-7923-3830-8
44. D.A. Hills, P.A. Kelly, D.N. Dai and A.M. Korsunsky: *Solution of Crack Problems*. The Distributed Dislocation Technique. 1996 ISBN 0-7923-3848-0
45. V.A. Squire, R.J. Hosking, A.D. Kerr and P.J. Langhorne: *Moving Loads on Ice Plates*. 1996 ISBN 0-7923-3953-3
46. A. Pineau and A. Zaoui (eds.): *IUTAM Symposium on Micromechanics of Plasticity and Damage of Multiphase Materials*. Proceedings of the IUTAM Symposium held in Sèvres, Paris, France. 1996 ISBN 0-7923-4188-0
47. A. Naess and S. Krenk (eds.): *IUTAM Symposium on Advances in Nonlinear Stochastic Mechanics*. Proceedings of the IUTAM Symposium held in Trondheim, Norway. 1996 ISBN 0-7923-4193-7
48. D. Ieşan and A. Scalia: *Thermoelastic Deformations*. 1996 ISBN 0-7923-4230-5

Mechanics

SOLID MECHANICS AND ITS APPLICATIONS

Series Editor: G.M.L. Gladwell

49. J.R. Willis (ed.): *IUTAM Symposium on Nonlinear Analysis of Fracture*. Proceedings of the IUTAM Symposium held in Cambridge, U.K. 1997 ISBN 0-7923-4378-6
50. A. Preumont: *Vibration Control of Active Structures*. An Introduction. 1997 ISBN 0-7923-4392-1
51. G.P. Cherepanov: *Methods of Fracture Mechanics: Solid Matter Physics*. 1997 ISBN 0-7923-4408-1
52. D.H. van Campen (ed.): *IUTAM Symposium on Interaction between Dynamics and Control in Advanced Mechanical Systems*. Proceedings of the IUTAM Symposium held in Eindhoven, The Netherlands. 1997 ISBN 0-7923-4429-4
53. N.A. Fleck and A.C.F. Cocks (eds.): *IUTAM Symposium on Mechanics of Granular and Porous Materials*. Proceedings of the IUTAM Symposium held in Cambridge, U.K. 1997 ISBN 0-7923-4553-3
54. J. Roorda and N.K. Srivastava (eds.): *Trends in Structural Mechanics*. Theory, Practice, Education. 1997 ISBN 0-7923-4603-3
55. Yu.A. Mitropolskii and N. Van Dao: *Applied Asymptotic Methods in Nonlinear Oscillations*. 1997 ISBN 0-7923-4605-X
56. C. Guedes Soares (ed.): *Probabilistic Methods for Structural Design*. 1997 ISBN 0-7923-4670-X
57. D. François, A. Pineau and A. Zaoui: *Mechanical Behaviour of Materials*. Volume I: Elasticity and Plasticity. 1998 ISBN 0-7923-4894-X
58. D. François, A. Pineau and A. Zaoui: *Mechanical Behaviour of Materials*. Volume II: Viscoplasticity, Damage, Fracture and Contact Mechanics. 1998 ISBN 0-7923-4895-8
59. L.T. Tenek and J. Argyris: *Finite Element Analysis for Composite Structures*. 1998 ISBN 0-7923-4899-0
60. Y.A. Bahei-El-Din and G.J. Dvorak (eds.): *IUTAM Symposium on Transformation Problems in Composite and Active Materials*. Proceedings of the IUTAM Symposium held in Cairo, Egypt. 1998 ISBN 0-7923-5122-3
61. I.G. Goryacheva: *Contact Mechanics in Tribology*. 1998 ISBN 0-7923-5257-2
62. O.T. Bruhns and E. Stein (eds.): *IUTAM Symposium on Micro- and Macrostructural Aspects of Thermoplasticity*. Proceedings of the IUTAM Symposium held in Bochum, Germany. 1999 ISBN 0-7923-5265-3
63. F.C. Moon: *IUTAM Symposium on New Applications of Nonlinear and Chaotic Dynamics in Mechanics*. Proceedings of the IUTAM Symposium held in Ithaca, NY, USA. 1998 ISBN 0-7923-5276-9
64. R. Wang: *IUTAM Symposium on Rheology of Bodies with Defects*. Proceedings of the IUTAM Symposium held in Beijing, China. 1999 ISBN 0-7923-5297-1
65. Yu.I. Dimitrienko: *Thermomechanics of Composites under High Temperatures*. 1999 ISBN 0-7923-4899-0
66. P. Argoul, M. Frémond and Q.S. Nguyen (eds.): *IUTAM Symposium on Variations of Domains and Free-Boundary Problems in Solid Mechanics*. Proceedings of the IUTAM Symposium held in Paris, France. 1999 ISBN 0-7923-5450-8
67. F.J. Fahy and W.G. Price (eds.): *IUTAM Symposium on Statistical Energy Analysis*. Proceedings of the IUTAM Symposium held in Southampton, U.K. 1999 ISBN 0-7923-5457-5
68. H.A. Mang and F.G. Rammerstorfer (eds.): *IUTAM Symposium on Discretization Methods in Structural Mechanics*. Proceedings of the IUTAM Symposium held in Vienna, Austria. 1999 ISBN 0-7923-5591-1

Mechanics

SOLID MECHANICS AND ITS APPLICATIONS

Series Editor: G.M.L. Gladwell

69. P. Pedersen and M.P. Bendsøe (eds.): *IUTAM Symposium on Synthesis in Bio Solid Mechanics*. Proceedings of the IUTAM Symposium held in Copenhagen, Denmark. 1999 ISBN 0-7923-5615-2
70. S.K. Agrawal and B.C. Fabien: *Optimization of Dynamic Systems*. 1999 ISBN 0-7923-5681-0
71. A. Carpinteri: *Nonlinear Crack Models for Nonmetallic Materials*. 1999 ISBN 0-7923-5750-7
72. F. Pfeifer (ed.): *IUTAM Symposium on Unilateral Multibody Contacts*. Proceedings of the IUTAM Symposium held in Munich, Germany. 1999 ISBN 0-7923-6030-3
73. E. Lavendelis and M. Zakrzhevsky (eds.): *IUTAM/IFToMM Symposium on Synthesis of Nonlinear Dynamical Systems*. Proceedings of the IUTAM/IFToMM Symposium held in Riga, Latvia. 2000 ISBN 0-7923-6106-7
74. J.-P. Merlet: *Parallel Robots*. 2000 ISBN 0-7923-6308-6
75. J.T. Pindera: *Techniques of Tomographic Isodyne Stress Analysis*. 2000 ISBN 0-7923-6388-4
76. G.A. Maugin, R. Drouot and F. Sidoroff (eds.): *Continuum Thermomechanics*. The Art and Science of Modelling Material Behaviour. 2000 ISBN 0-7923-6407-4
77. N. Van Dao and E.J. Kreuzer (eds.): *IUTAM Symposium on Recent Developments in Non-linear Oscillations of Mechanical Systems*. 2000 ISBN 0-7923-6470-8
78. S.D. Akbarov and A.N. Guz: *Mechanics of Curved Composites*. 2000 ISBN 0-7923-6477-5
79. M.B. Rubin: *Cosserat Theories: Shells, Rods and Points*. 2000 ISBN 0-7923-6489-9
80. S. Pellegrino and S.D. Guest (eds.): *IUTAM-IASS Symposium on Deployable Structures: Theory and Applications*. Proceedings of the IUTAM-IASS Symposium held in Cambridge, U.K., 6–9 September 1998. 2000 ISBN 0-7923-6516-X
81. A.D. Rosato and D.L. Blackmore (eds.): *IUTAM Symposium on Segregation in Granular Flows*. Proceedings of the IUTAM Symposium held in Cape May, NJ, U.S.A., June 5–10, 1999. 2000 ISBN 0-7923-6547-X
82. A. Lagarde (ed.): *IUTAM Symposium on Advanced Optical Methods and Applications in Solid Mechanics*. Proceedings of the IUTAM Symposium held in Futuroscope, Poitiers, France, August 31–September 4, 1998. 2000 ISBN 0-7923-6604-2
83. D. Weichert and G. Maier (eds.): *Inelastic Analysis of Structures under Variable Loads*. Theory and Engineering Applications. 2000 ISBN 0-7923-6645-X
84. T.-J. Chuang and J.W. Rudnicki (eds.): *Multiscale Deformation and Fracture in Materials and Structures*. The James R. Rice 60th Anniversary Volume. 2001 ISBN 0-7923-6718-9
85. S. Narayanan and R.N. Iyengar (eds.): *IUTAM Symposium on Nonlinearity and Stochastic Structural Dynamics*. Proceedings of the IUTAM Symposium held in Madras, Chennai, India, 4–8 January 1999 ISBN 0-7923-6733-2
86. S. Murakami and N. Ohno (eds.): *IUTAM Symposium on Creep in Structures*. Proceedings of the IUTAM Symposium held in Nagoya, Japan, 3-7 April 2000. 2001 ISBN 0-7923-6737-5
87. W. Ehlers (ed.): *IUTAM Symposium on Theoretical and Numerical Methods in Continuum Mechanics of Porous Materials*. Proceedings of the IUTAM Symposium held at the University of Stuttgart, Germany, September 5-10, 1999. 2001 ISBN 0-7923-6766-9
88. D. Durban, D. Givoli and J.G. Simmonds (eds.): *Advances in the Mechanis of Plates and Shells The Avinoam Libai Anniversary Volume*. 2001 ISBN 0-7923-6785-5
89. U. Gabbert and H.-S. Tzou (eds.): *IUTAM Symposium on Smart Structures and Structonic Systems*. Proceedings of the IUTAM Symposium held in Magdeburg, Germany, 26–29 September 2000. 2001 ISBN 0-7923-6968-8

Mechanics

SOLID MECHANICS AND ITS APPLICATIONS

Series Editor: G.M.L. Gladwell

90. Y. Ivanov, V. Cheshkov and M. Natova: *Polymer Composite Materials – Interface Phenomena & Processes*. 2001 ISBN 0-7923-7008-2
91. R.C. McPhedran, L.C. Botten and N.A. Nicorovici (eds.): *IUTAM Symposium on Mechanical and Electromagnetic Waves in Structured Media*. Proceedings of the IUTAM Symposium held in Sydney, NSW, Australia, 18-22 Januari 1999. 2001 ISBN 0-7923-7038-4
92. D.A. Sotiropoulos (ed.): *IUTAM Symposium on Mechanical Waves for Composite Structures Characterization*. Proceedings of the IUTAM Symposium held in Chania, Crete, Greece, June 14-17, 2000. 2001 ISBN 0-7923-7164-X
93. V.M. Alexandrov and D.A. Pozharskii: *Three-Dimensional Contact Problems*. 2001 ISBN 0-7923-7165-8
94. J.P. Dempsey and H.H. Shen (eds.): *IUTAM Symposium on Scaling Laws in Ice Mechanics and Ice Dynamics*. Proceedings of the IUTAM Symposium held in Fairbanks, Alaska, U.S.A., 13-16 June 2000. 2001 ISBN 1-4020-0171-1
95. U. Kirsch: *Design-Oriented Analysis of Structures*. A Unified Approach. 2002 ISBN 1-4020-0443-5
96. A. Preumont: *Vibration Control of Active Structures*. An Introduction (2nd Edition). 2002 ISBN 1-4020-0496-6
97. B.L. Karihaloo (ed.): *IUTAM Symposium on Analytical and Computational Fracture Mechanics of Non-Homogeneous Materials*. Proceedings of the IUTAM Symposium held in Cardiff, U.K., 18-22 June 2001. 2002 ISBN 1-4020-0510-5
98. S.M. Han and H. Benaroya: *Nonlinear and Stochastic Dynamics of Compliant Offshore Structures*. 2002 ISBN 1-4020-0573-3
99. A.M. Linkov: *Boundary Integral Equations in Elasticity Theory*. 2002 ISBN 1-4020-0574-1
100. L.P. Lebedev, I.I. Vorovich and G.M.L. Gladwell: *Functional Analysis*. Applications in Mechanics and Inverse Problems (2nd Edition). 2002 ISBN 1-4020-0667-5; Pb: 1-4020-0756-6
101. Q.P. Sun (ed.): *IUTAM Symposium on Mechanics of Martensitic Phase Transformation in Solids*. Proceedings of the IUTAM Symposium held in Hong Kong, China, 11-15 June 2001. 2002 ISBN 1-4020-0741-8
102. M.L. Munjal (ed.): *IUTAM Symposium on Designing for Quietness*. Proceedings of the IUTAM Symposium held in Bangkok, India, 12-14 December 2000. 2002 ISBN 1-4020-0765-5
103. J.A.C. Martins and M.D.P. Monteiro Marques (eds.): *Contact Mechanics*. Proceedings of the 3rd Contact Mechanics International Symposium, Praia da Consolação, Peniche, Portugal, 17-21 June 2001. 2002 ISBN 1-4020-0811-2
104. H.R. Drew and S. Pellegrino (eds.): *New Approaches to Structural Mechanics, Shells and Biological Structures*. 2002 ISBN 1-4020-0862-7
105. J.R. Vinson and R.L. Sierakowski: *The Behavior of Structures Composed of Composite Materials*. Second Edition. 2002 ISBN 1-4020-0904-6
106. Not yet published.
107. J.R. Barber: *Elasticity*. Second Edition. 2002 ISBN Hb 1-4020-0964-X; Pb 1-4020-0966-6
108. C. Miehe (ed.): *IUTAM Symposium on Computational Mechanics of Solid Materials at Large Strains*. Proceedings of the IUTAM Symposium held in Stuttgart, Germany, 20-24 August 2001. 2003 ISBN 1-4020-1170-9

Mechanics

SOLID MECHANICS AND ITS APPLICATIONS

Series Editor: G.M.L. Gladwell

109. P. Ståhle and K.G. Sundin (eds.): *IUTAM Symposium on Field Analyses for Determination of Material Parameters – Experimental and Numerical Aspects.* Proceedings of the IUTAM Symposium held in Abisko National Park, Kiruna, Sweden, July 31 – August 4, 2000. 2003 ISBN 1-4020-1283-7
110. N. Sri Namachchivaya and Y.K. Lin (eds.): *IUTAM Symposium on Nonlinear Stochastic Dynamics.* Proceedings of the IUTAM Symposium held in Monticello, IL, USA, 26 – 30 August, 2000. 2003 ISBN 1-4020-1471-6
111. H. Sobieckzky (ed.): *IUTAM Symposium Transsonicum IV.* Proceedings of the IUTAM Symposium held in Göttingen, Germany, 2–6 September 2002, 2003 ISBN 1-4020-1608-5
112. J.-C. Samin and P. Fisette: *Symbolic Modeling of Multibody Systems.* 2003 ISBN 1-4020-1629-8
113. A.B. Movchan (ed.): *IUTAM Symposium on Asymptotics, Singularities and Homogenisation in Problems of Mechanics.* Proceedings of the IUTAM Symposium held in Liverpool, United Kingdom, 8-11 July 2002. 2003 ISBN 1-4020-1780-4
114. S. Ahzi, M. Cherkaoui, M.A. Khaleel, H.M. Zbib, M.A. Zikry and B. LaMatina (eds.): *IUTAM Symposium on Multiscale Modeling and Characterization of Elastic-Inelastic Behavior of Engineering Materials.* Proceedings of the IUTAM Symposium held in Marrakech, Morocco, 20-25 October 2002. 2004 ISBN 1-4020-1861-4
115. H. Kitagawa and Y. Shibutani (eds.): *IUTAM Symposium on Mesoscopic Dynamics of Fracture Process and Materials Strength.* Proceedings of the IUTAM Symposium held in Osaka, Japan, 6-11 July 2003. Volume in celebration of Professor Kitagawa's retirement. 2004 ISBN 1-4020-2037-6
116. E.H. Dowell, R.L. Clark, D. Cox, H.C. Curtiss, Jr., K.C. Hall, D.A. Peters, R.H. Scanlan, E. Simiu, F. Sisto and D. Tang: *A Modern Course in Aeroelasticity.* 4th Edition, 2004 ISBN 1-4020-2039-2
117. T. Burczyński and A. Osyczka (eds.): *IUTAM Symposium on Evolutionary Methods in Mechanics.* Proceedings of the IUTAM Symposium held in Cracow, Poland, 24-27 September 2002. 2004 ISBN 1-4020-2266-2
118. D. Ieşan: *Thermoelastic Models of Continua.* 2004 ISBN 1-4020-2309-X
119. G.M.L. Gladwell: *Inverse Problems in Vibration.* Second Edition. 2004 ISBN 1-4020-2670-6
120. J.R. Vinson: *Plate and Panel Structures of Isotropic, Composite and Piezoelectric Materials, Including Sandwich Construction.* 2005 ISBN 1-4020-3110-6
121. *Forthcoming*
122. G. Rega and F. Vestroni (eds.): *IUTAM Symposium on Chaotic Dynamics and Control of Systems and Processes in Mechanics.* Proceedings of the IUTAM Symposium held in Rome, Italy, 8–13 June 2003. 2005 ISBN 1-4020-3267-6
123. E.E. Gdoutos: *Fracture Mechanics. An Introduction. 2nd edition.* 2005 ISBN 1-4020-3267-6
124. M.D. Gilchrist (ed.): *IUTAM Symposium on Impact Biomechanics from Fundamental Insights to Applications.* 2005 ISBN 1-4020-3795-3
125. J.M. Huyghe, P.A.C. Raats and S. C. Cowin (eds.): *IUTAM Symposium on Physicochemical and Electromechanical Interactions in Porous Media.* 2005 ISBN 1-4020-3864-X
126. H. Ding, W. Chen and L. Zhang: *Elasticity of Transversely Isotropic Materials.* 2005 ISBN 1-4020-4033-4
127. W. Yang (ed): *IUTAM Symposium on Mechanics and Reliability of Actuating Materials.* Proceedings of the IUTAM Symposium held in Beijing, China, 1–3 September 2004. 2005 ISBN 1-4020-4131-6

Mechanics

SOLID MECHANICS AND ITS APPLICATIONS

Series Editor: G.M.L. Gladwell

128. J.-P. Merlet: *Parallel Robots.* 2006 ISBN 1-4020-4132-2
129. G.E.A. Meier and K.R. Sreenivasan (eds.): *IUTAM Symposium on One Hundred Years of Boundary Layer Research.* Proceedings of the IUTAM Symposium held at DLR-Göttingen, Germany, August 12–14, 2004. 2006 ISBN 1-4020-4149-7
130. H. Ulbrich and W. Günthner (eds.): *IUTAM Symposium on Vibration Control of Nonlinear Mechanisms and Structures.* 2006 ISBN 1-4020-4160-8
131. L. Librescu and O. Song: *Thin-Walled Composite Beams.* Theory and Application. 2006 ISBN 1-4020-3457-1
132. G. Ben-Dor, A. Dubinsky and T. Elperin: *Applied High-Speed Plate Penetration Dynamics.* 2006 ISBN 1-4020-3452-0
133. X. Markenscoff and A. Gupta (eds.): *Collected Works of J. D. Eshelby.* Mechanics and Defects and Heterogeneities. 2006 ISBN 1-4020-4416-X
134. R.W. Snidle and H.P. Evans (eds.): *IUTAM Symposium on Elastohydrodynamics and Microelastohydrodynamics.* Proceedings of the IUTAM Symposium held in Cardiff, UK, 1–3 September, 2004. 2006 ISBN 1-4020-4532-8
135. T. Sadowski (ed.): *IUTAM Symposium on Multiscale Modelling of Damage and Fracture Processes in Composite Materials.* Proceedings of the IUTAM Symposium held in Kazimierz Dolny, Poland, 23–27 May 2005. 2006 ISBN 1-4020-4565-4
136. A. Preumont: *Mechatronics.* Dynamics of Electromechanical and Piezoelectric Systems. 2006 ISBN 1-4020-4695-2
137. M.P. Bendsøe, N. Olhoff and O. Sigmund (eds.): *IUTAM Symposium on Topological Design Optimization of Structures, Machines and Materials.* Status and Perspectives. 2006 ISBN 1-4020-4729-0
138. A. Klarbring: *Models of Mechanics.* 2006 ISBN 1-4020-4834-3
139. H.D. Bui: *Fracture Mechanics.* Inverse Problems and Solutions. 2006 ISBN 1-4020-4836-X
140. M. Pandey, W.-C. Xie and L. Xu (eds.): *Advances in Engineering Structures, Mechanics and Construction.* Proceedings of an International Conference on Advances in Engineering Structures, Mechanics & Construction, held in Waterloo, Ontario, Canada, May 14–17, 2006. 2006 ISBN 1-4020-4890-4
141. G.Q. Zhang, W.D. van Driel and X. J. Fan: *Mechanics of Microelectronics.* 2006 ISBN 1-4020-4934-X
142. Q.P. Sun and P. Tong (eds.): *IUTAM Symposium on Size Effects on Material and Structural Behavior at Micron- and Nano-Scales.* Proceedings of the IUTAM Symposium held in Hong Kong, China, 31 May–4 June, 2004. 2006 ISBN 1-4020-4945-5

MIX
Papier aus verantwortungsvollen Quellen
Paper from responsible sources
FSC® C105338
FSC
www.fsc.org

If you have any concerns about our products,
you can contact us on
ProductSafety@springernature.com

In case Publisher is established outside the EU,
the EU authorized representative is:
Springer Nature Customer Service Center GmbH
Europaplatz 3, 69115 Heidelberg, Germany

Printed by Libri Plureos GmbH
in Hamburg, Germany